Separation of Gases

Special Publication No. 80

Separation of Gases

The Proceedings of the Fifth BOC Priestley Conference,
sponsored by BOC Limited and organised by the Royal Society of
Chemistry in conjunction with the University of Birmingham

Birmingham, 19th–21st September 1989

ROYAL
SOCIETY OF
CHEMISTRY

British Library Cataloguing in Publication Data

BOC Priestley Conference (5th :1989: Birmingham),
   England
   Separation of gases.
   1. Gases. Separation
   I. Title   II. Royal Society of Chemistry   III. University
   of Birmingham   IV. Series
   660.2842

ISBN 0–85186–637–9

Published by the Royal Society of Chemistry, Thomas Graham House, Science Park,
Cambridge CB4 4WF

Printed in Great Britain by Henry Ling Ltd., at the Dorset Press, Dorchester, Dorset

Preface

This fifth BOC Priestley Conference held at the University of Birmingham followed a similar pattern to previous conferences in the series. The programme embraced many of the diverse scientific and religious interests of Joseph Priestley (1733-1804) who discovered oxygen and investigated the properties of a number of gases, in addition to his occupation as a preacher and minister of the Unitarian Church.

The scientific component of the programme for the fifth BOC Priestley Conference was devoted to the separation of gases, a topic which currently offers a challenge to the intellect and skills of scientists, engineers and industrialists. The economics and eventual commercial success of many process operations depends crucially on the ease with which products of reaction can be separated in specific plant located downstream from the chemical reactor. While distillation and absorption remain well established, and, in many circumstances, convenient and appropriate methods of separation, other methods of separation have begun to take precedence where it is perceived that an economic advantage accrues. Most of these more recently developed techniques for large-scale gas separation were included as part of the conference scientific programme, which included invited lectures on separation by membranes, adsorption, high gravity separations and chemical methods of separation. Both fundamental scientific aspects of gas separation and the challenging intellectual aspects of process design were included in the programme.

Because one of the objectives of the BOC Priestley Conference is to commemorate the life and work of Priestley, it is apposite that one session is devoted to historical matters surrounding Priestley's work. Four distinguished speakers have selected subjects which relate to the early scientific practices and problems which emerged during the industrial revolution when new processes, many of which were based on the discoveries and ideas of Priestley, began to burgeon.

Priestley's life as a philosopher and spiritual leader in the communities of both Leeds and Birmingham culminated in rather tragic circumstances. Nevertheless his contribution to religious thought has stimulated theologians to many avenues of thought. The Priestley lecture by the Reverend Dr. J. Polkinghorne, FRS, entitled 'A Scientist's View of Religion', therefore provides us with the opportunity of reviewing our own individual experiences of life and our attitude towards spiritual matters.

The BOC Centenary Lecture, sponsored by BOC, was delivered on this occasion by Emeritus Professor G.G. Haselden, who chose as his subject 'Gas Separation Fundamentals'. No topic could have been more appropriate for this conference than fundamental issues relating to gas separation, because whether one is concerned with advancing scientific knowledge or accepting the challenge of designing new process engineering routes, it is the fundamental development of scientific and engineering ideas that takes us further forward.

W.J. Thomas
Chairman
5th BOC Priestley
Programme Committee

Contents

BOC Centenary Lecture 1989

Gas Separation Fundamentals

G. G. Haselden

DEPARTMENT OF CHEMICAL ENGINEERING, UNIVERSITY OF LEEDS, LEEDS LS2 9JT, UK

1 INTRODUCTION

The aim of this lecture is to make an up-to-date engineering
assessment of industrial gas separation. The word 'engineering'
is inserted deliberately and implies going beyond a scientific
account of the virtues of rival processes to considerations of
cost. This extension does not necessarily imply a departure from
fundamentals; in fact we shall be concerned to discover how far
thermodynamics, molecular kinetics and other basic sciences govern
the overall attractiveness of processes.

Before tackling costs two issues must be faced, the first is
the full specification of the separation task to be performed, and
the second is safety.

For a given feed gas mixture, and flowrate, the specification
must cover the purity of the various product fractions, and the
amount of any residue. There may be a need for turn-down, if
demand is not steady, whilst in other cases the emphasis may be on
achieving long periods of uninterrupted operation. The pressure
at which the feed gas is available, and that at which one or more
of the products is required, may greatly influence process
selection. For instance natural gas may be available at 100 bar
whilst air is normally available at atmospheric pressure.

The safety requirement is paramount, but decisions are often
complex. In the emerging 'green' culture there a prevailing
mood which classifies processes as either safe or unsafe with no
acceptable ground between them. Taking twenty tons of petrol
through a town centre in a thin-walled tank is acceptable, whilst
conveying radio-active material by rail in containers designed to
withstand high velocity impact is unacceptable. The trouble
surfaces if risks have to be quantified. Understandably nobody

wants to put a number to an acceptable risk.   So there is a strong
temptation to management to play down any potential hazard rather
than offering it for rational discussion.

In gas separation terms safety questions will arise when one
or more of the components is flammable or toxic.   Then a
separation process operating at atmospheric pressure is likely to
be inherently safer than one operating at high pressure – or even
under modest vacuum.   But a high pressure process may well be much
cheaper and use less power.   Which should be used?

The ideal answer is a wise government which lays down and
enforces codes of design and practice which eliminate known risks,
and require a high degree of integrity in tackling new ones.   In
the absence of this panacea management must ensure that the
economic pointer is followed only when all risks have been
thoroughly explored and answered responsibly.

2 OPERATING COSTS

The direct operating costs of a gas separation plant normally
involve at least three elements:

(i)   Cost of feedstock.

Here there are exceptions, like air separation or flue gas
clean-up, where the feed is free.   Where the feed must be paid for
economic pressures will require that the minimum is wasted as
residue.

(ii) Running costs (such as fuel, electricity, labour,
maintenance, etc.)

Frequently the fuel and energy bill is dominant; then
thermodynamics comes into its own.   Entropy costs money, and as
little as possible should be created.   Increasingly separation
processes are being subjected to thermodynamic analysis using
techniques such as pinch-technology (1). Computer based control is
being adopted increasingly to reduce both wastage and operating
labour costs.

(iii) Capital charges (comprising amortization and interest
payments).

With any separation process the cheapest plant is likely to
have the highest running cost, and a balance must be struck.
Inflation, market uncertainties, and high interest rates tempt
management to take a short-term view, whilst increasing power costs
reward those taking the longer view.

The relative capital costs of different separation processes are determined in some measure by molecular kinetics. The faster the inherent separation mechanism the smaller will be the volume of plant required to contain it, and the lower its cost. Thus distillation columns are relatively compact for their throughput because the molecular processes at a liquid/vapour interface are very rapid. Absorption and adsorption are slower because they involve diffusion through stationery phases. Membrane processes tend to be even slower because they involve diffusional resistances in both the gas and membrane phases.

Capital charges must relate to the installed and commissioned plant. For a plant erected on site the final cost will normally be about four times the total cost of all the major plant components. In this respect packaged plants can reduce first costs as well as reducing uncertainties.

In summary, the best plant for a given separation duty is one which fully matches specification, is safe, and has the lowest overall operating cost (obtained by minimizing the sum of running and capital charges).

3 MINIMUM WORK OF SEPARATION OF A GAS MIXTURE

The thermodynamic minimum work of separation of a gas mixture is readily calculated from the free energy difference between the gas feed and the total products. For gases having a negligible heat of mixing, and at pressures well below the critical, ideal separation can be modelled as in Fig.1. The figure is for a binary mixture, but is easily extended for complex mixtures. It presupposes the existence of 2 different membranes, one perfectly permeable to species a and impermeable to species b, and the second having the converse properties. The upper piston is free to move downwards to maintain the feed at constant pressure which will be taken as atmospheric pressure. The gas a passes through its membrane at its partial pressure P_a and accumulates in the left-hand cylinder as the corresponding piston moves down. Simultaneously the piston in the right-hand cylinder moves down so as to maintain the feed at constant composition. When the pistons have reached the bottom of their travel the motion is reversed, the non-return valves communicating with the membrane chambers close, and the gases start to compress. When the pressure of each product reaches atmospheric pressure the discharge valves open and the pure gases are expelled at constant pressure. For the minimum work requirement the gases must be compressed and discharged isothermally and reversibly. Hence the minimum molar work of separation, at temperature T and pressure P, and for mole fractions y, is given by:

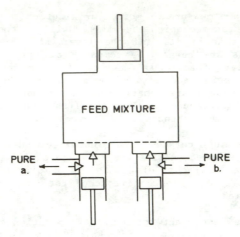

<u>Figure 1</u> Model for the ideal separation of a binary gas mixture using perfect membranes

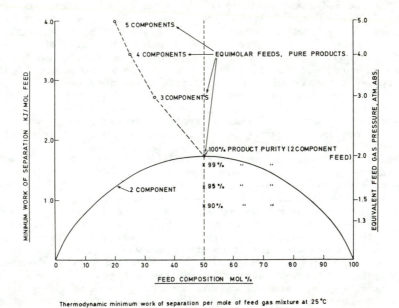

Thermodynamic minimum work of separation per mole of feed gas mixture at 25 °C

<u>Figure 2</u> The minimum work of separation of a range of different gas mixtures

$$W_{min} = RT \left[y_a \ln \frac{P}{P_a} + y_b \ln \frac{P}{P_b} \right]$$

$$= RT \left[y_a \ln \frac{1}{y_a} + y_b \ln \frac{1}{y_b} \right]$$

The symmetrical smooth curve in Fig. 2 shows the thermodynamic minimum work of separation of a two-component mixture as a function of feed composition. The greatest work of about 1.72 kJ/mol is required for an equimolar mixture, whilst a mixture such as air (treated as a binary mixture with 21% oxygen) requires about 1.3 kJ/mol. The further points on the centre-line give the minimum work requirements for separating an equimolar feed into less pure products. The effect of purity is substantial, 90% products requiring only about half the energy of separation of pure products.

In the upper part of the diagram are shown the minimum separation work requirements of multicomponent mixtures. On the right-hand axis of the diagram is shown the equivalent feed gas pressure if the minimum work of separation is expended in isothermally compressing it. For instance an ideal air separation plant (requiring 1.3 kJ/mol for the binary separation) would require an air feed pressure of only about 1.7 atm. abs. It is interesting to note that for multicomponent mixtures the required feed pressure in atmospheres is precisely equal to the number of components - a fact which is readily understood by pondering Fig. 1.

4 CRYOGENIC DISTILLATION

The most important field of industrial gas separation - air separation - is dominated by cryogenic distillation. This may seem strange in view of the need to cool air to about 90K to liquefy it. But this hurdle is more than compensated in most applications by the following advantages:

(i) Highly compact plant

The rapid molecular movements at the vapour/liquid interface have already been referred to. It means that in the lower column of a modern air separation plant each cubic meter of column separates more than 100 tonnes of air per day. Also its construction is quite cheap.

(ii) Readily staged

Distillation columns can be designed to match precisely the separation required, yielding multiple products of differing purity.

(iii) Continuous operation

Steady flow makes for ease of control, and for long life of plant components.

To separate a binary mixture by distillation using the thermodynamic minimum work requires a distillation system as shown in Fig. 3.   The feed gas, cooled to its dew point enters the central region of a column having an infinite number of plates. At every plate below the feed there is an incremental addition of heat at progressively increasing temperature so that the reboil ratio is kept precisely at the minimum.   This is done by heat exchange with a circulating pure fluid which progressively condenses at gradually reducing pressures produced by a series of ideal expanders.   Similarly above the feed the reflux ratio is gradually raised (though always at the minimum value) by the progressive evaporation, at diminishing pressure, of the circulated fluid.   This fluid finally emerges as vapour, and is compressed isentropically to its dew point pressure at the reboiler temperature.   It has been shown that this compression work, minus the sum of all the incremental work quantities recovered in the expanders, is precisely equal to the thermodynamic minimum work of separation.

As usual the ideal model is far from practical.   In particular there is great simplification if the separation work can be applied by compressing the feed gas rather than a recirculated fluid.   Also a cryogenic distillation plant must include two other provisions.   It must provide purification means to eliminate harmful contaminants such as water vapour, carbon dioxide and hydrocarbons from the feed air.   It must also incorporate refrigeration provision to cool the plant initially and to compensate for steady-state cold losses.

Fig. 4 shows the simplified flow sheet of a modern oxygen plant described by Wilson et al (2), based on the Linde double column principle, which requires a feed air pressure of 6.6 bar. The air feed is purified by adsorption (not shown) and refrigeration is generated by the air expansion turbine fed by an intermediate take-off from the heat exchanger.   The reversibility of the double column is improved by condensing part of the feed air in the oxygen product vaporizer and using it as an additional reflux to both columns.   The theoretical work requirement of this separation is complicated by the fact that some high purity nitrogen is produced at a modest pressure and the argon product

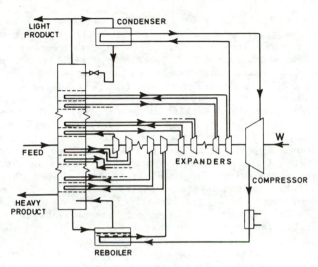

<u>Figure 3</u> Thermodynamic model of an ideal distillation column.

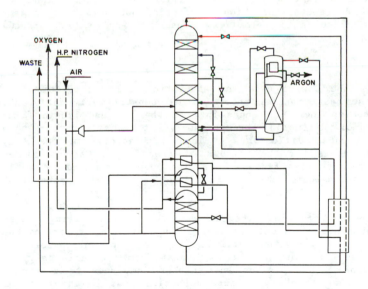

<u>Figure 4</u> The flowsheet of a modern air separation plant

leaves as liquid.   On this basis the overall thermodynamic
efficiency of the plant is calculated to be 26.1% - a high value.
The wasted energy is largely accounted for by the compressor (30%
of losses), heat exchangers (30%) and the distillation column
(22%).   Whilst small improvements of compressors and heat
exchangers can still be looked for the fight against the law of
diminishing returns is already present.   There is scope for
further improvement of the distillation system, especially when one
or more of the products is required at a purity not greater than
98%.   This may be done by thermally linking the high and low
pressure columns, a pursuit in which I was engaged for several
years (3). Fig. 5 shows one possible method which has been studied
in the Chemical Engineering Department at Leeds. Wilson describes
a partially linked column system (Fig.6) developed for medium
purity oxygen in which the separation efficiency has been raised to
nearly 30%.   This flowsheet brings out the advantages of the
cryogenic distillation approach in that all the energy consumption
occurs in a single large compressor, all the components operate at
high mass flux densities, and the complexity is in the plumbing
rather than the components. Thus high efficiency can be associated
with relatively low investment cost.

        I believe that cryogenic distillation will continue to
dominate those separations which are carried out on a high tonnage
scale, for mixtures of relatively close boiling ($< 20°C$ spread)
components.

        5 ABSORPTION

        Whilst the towers used for absorption and distillation are
similar, the mass transfer processes are substantially different.
With absorption the resistances in both the gas and liquid films
are those of diffusion through an inert phase, rather than
equimolar counter diffusion.   Thus for absorption the tray
efficiency values (or the heights of transfer units for packed
columns) are much less favourable. The compensating advantage is
that separation can often be effected at a more convenient
temperature.

        Absorption comes into its own when the normal boiling points
of the components are widely separated, or where one or more of the
components have a strong affinity for a particular solvent.
Hence its use with carbon dioxide removal from synthesis gas, and
for scrubbing carbon dioxide and sulphur compounds from natural
gas.   The absorption may be purely physical, as in the Rectisol
process using cold methanol, or largely chemical as in hot
potassium carbonate scrubbing (Benfield and similar processes), or
a combination of the two.   Much development in recent years has
centred on tailoring solvents to achieve selectivity to the
required components (4).

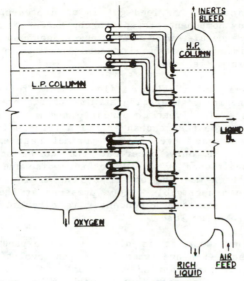

<u>Figure 5</u> A method of thermally linking two columns so that the reflux ratio in both is kept close to the minimum

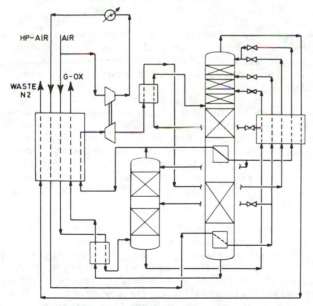

<u>Figure 6</u> The flowsheet of a medium purity oxygen plant in which some variation of reflux and reboil is achieved

In many applications the feed gas is at a pressure of 20 bar or more so that any component present to the extent of more than 5% is potentially removable with zero power consumption, whilst leaving the remaining gas at feed pressure. In practice this possibility is seldom realised.

Fig. 7 gives an idealized flow sheet for pressurized absorption with thermal regeneration at atmospheric pressure. The large mass flowrate of solvent compared to the mass of components absorbed means that the heat exchanger has a large effect on steam and cooling water usage. Even with a perfect heat exchanger the temperature pinch can only occur at the cold end, and this temperature is conditioned by the temperature rise due to the heat of absorption - which must be offset by the cooler. An ideal solvent would have a negligible vapour pressure at the top of the regenerator but, in practice a condenser is often necessary, and this adds to the reboiler load.

To reduce pump work it is desirable to expand the loaded solvent after heat exchange when some gas evolution will be present, and will continue during expansion. This duty will not be handled readily by a turbine, but it might be handled by a screw compressor used in reverse. If there is a significant amount of gas evolution the combined expander and pump should generate nett shaft work.

Ideally the steam requirement in the reboiler is only that required to desorb the acid gases not evolved in the heat exchanger and expander. In fact steam consumption is far greater due to the heat exchanger inefficiency, volatilization of the solvent, and other heat losses.

The recently announced MOLTOX (5) process uses absorption in a molten mixture of alkali (Na/K) nitrates and nitrites to separate oxygen from air in the temperature range 500 to 650°C. The chemical reaction from nitrite to nitrate, by which the oxygen is taken up, is promoted by a soluble catalyst, but it requires that the feed air is dry and CO_2-free. The oxygen loaded molten salt may be regenerated either by dropping the pressure or raising the temperature.

Because the absorption process operates at temperatures so far above ambient (with considerable associated costs) it is not advocated as a free-standing oxygen process. Rather it is coupled with power generation or with the supply of high grade heat to a suitable metallurgical or chemical process. Since the oxygen process consumes very little heat, but only downgrades it from say 680° to 540°C, the cost of the oxygen thus produced depends critically on the method of assessing its share of fuel and energy costs, and whether it bears any capital charges associated with purifying and heating the feed air and cooling the products.

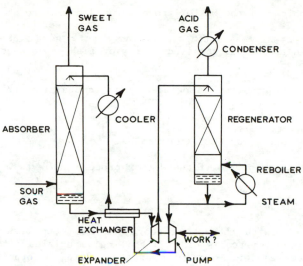

Figure 7 The flowsheet of an idealized absorption cycle with thermal regeneration

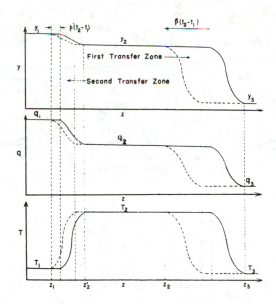

Figure 8 Concentration and temperature profiles for adiabatic adsorption of a non-dilute gas feed into a regenerated bed, at two different time intervals

    The problems of the MOLTOX process would appear to be:

(i)   Achieving a close match at all times between the utilisation
of oxygen and the demand for electricity or high grade heat for the
associated plant.

(ii)   Materials problems associated with containing the molten
salt.

(iii) As the thermodynamic efficiency of the associated plant
increases the scope for MOLTOX decreases.

    I shall be surprised if the Moltox process achieves widespread
adoption.

    6 ADSORPTION

    Adsorption with thermal regeneration has long been a sound
method for removing traces of high boiling impurities from gases.
Its much wider adoption came with pressure swing adsorption (PSA)
which avoids the inherently slow and irreversible process of
heating all the adsorbent, and its container, to the regeneration
temperature.   The best adsorbent for use in PSA should have a high
adsorptivity for at least one of the components.   To facilitate
regeneration this absorptivity should be a strong and reversible
function of pressure and temperature.   The adsorbent also requires
to be physically strong and inexpensive.

    PSA involves the passage of concentration and temperature
fronts through fixed beds of adsorbent during both adsorption and
regeneration (6).   An example, for the case of adsorption, is
presented in Fig. 8 by Pan and Basmadjian (7).   The dotted
profiles correspond to a short time interval t_1, and the full lines
to a longer interval t_2.   All the profiles exhibit two transfer
zones because the heat of absorption liberated initially raises the
bed temperature, thus reducing its adsorption potential.   When
more feed gas has passed through this part of the bed its
temperature will drop and adsorption can rise to the equilibrium
level.

    Preferably the concentration and temperature profiles should
both be steep and nearly coincident.   Shallow profiles imply low
overall bed capacity and large dilution losses at change-over.
Steep profiles require favourable adsorption isotherms together
with low heat and mass transfer resistances within and around the
adsorbent particles.   Furthermore the bed must be packed uniformly
and be free of wall effects.   With these provisions the passage of
the profiles through the adsorbent bed will approach thermodynamic
reversibility.

    The main losses with PSA occur during regeneration.   To

reduce these losses a wide, and somewhat confusing, range of
multi-bed systems has been developed. A very useful summary is
furnished by Yang (6), whilst White and Barkley (8) outline present
design practice.

The 4-bed adsorption system (Fig. 9), as described by Doshi et
al (9), will suffice for the present analysis. The time cycle is
divided into 4 equal periods, but these are sometimes sub-divided.
We will follow the cycle of bed A which starts already pressurized,
and performs adsorption throughout the first period until the
break-through curves approach the bed outlet. At this instant
vessel A contains gas at feed composition over most of its length,
together with gas of product composition over a short length
adjacent to the product end, and gas of intermediate composition
over a comparatively short transition zone. All this gas is at
near feed pressure and, in total quantity, is comparable with the
amount of gas adsorbed. The challenge is to recover the gas in
these different zones without mixing.

In the second period bed A has become bed C. At the beginning
of the period the discharge ends of beds C and D are interconnected
and pressure equalisation occurs. Two benefits accrue; the H.P.
product gas near the discharge end of C is trapped in D (which had
been fully regenerated in the previous stage), and part of the
pressure energy is recovered. Later in the second period the
discharge end of C is connected through a pressure let-down valve
to the discharge end of B so that the resulting purge flow through
B displaces much of the absorbate (L.P. product) and supplies the
necessary heat of desorption.

In the third time interval bed C has become bed B. Firstly it
experiences 'dump', when the inlet end is directly connected to the
L.P. product line. Some adsorbate is desorbed and the bed
temperature drops. Then B receives the purge flow from C and the
breakthrough curves make their reverse passage through the bed
until it is almost completely regenerated.

During the final time interval bed B has become bed D. It is
partially pressurized with H.P. product from C, and then totally
pressurized with H.P. product. Thus the full cycle is completed.

The sources of purity and energy losses in the total cycle are
numerous and interrelated. Pressure let-down is irreversible and
leads to local gradients of both composition and temperature.
These are compounded by end effects due to headers, pipelines and
valves. Multiple let-down stages reduce the irreversibility and
are justified for large throughputs. Plants have been built with
as many as ten interconnected beds.

Some PSA systems use vacuum in regeneration, whilst others
employ 'rinse' stages in which the residual gas in a bed is

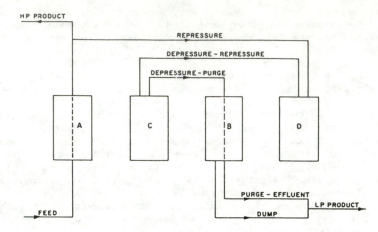

Figure 9 A 4-bed PSA system

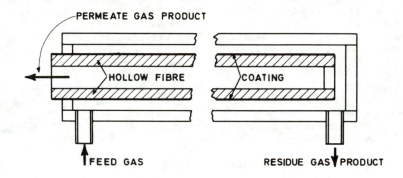

Figure 10 A simplified hollow fibre membrane separator, and its use
          in natural gas sweetening.

	FEED GAS	PERMEATE	RESIDUE
Pressure (bar)	60.0	1.7	58.8
% CH_4	93.0	63.4	98.0
% CO_2	7.0	36.6	2.0
Partial Pressure CO_2 (bar)	4.2	0.6	1.2
Quantity %	100	15	85

displaced co-currently at feed pressure by a stream of adsorbate. Yet other processes use pressure equalization tanks to buffer pressure changes, or mix high heat capacity inert solid with the adsorbent to buffer temperature swing.   There is scope for an expert system to guide the designer through this maze.

Optimum bed design starts with the size and shape of adsorbent particles to achieve rapid mass transfer without unacceptable pressure loss.   The beds should be as short as possible without incurring excessive end losses on reversal.   Practical challenges arise in minimising wall losses, and in filtering the feed gas to prevent clogging of the adsorbent pores with particulates.

It is reasonable to expect a considerable growth of PSA.   The underlying molecular processes are rapid and reasonably reversible, and control is effected readily by computer.

7 MEMBRANES

The self evident simplicity of membrane separation is highly attractive, but perhaps misleading.   Much excellent work has been done in developing new membranes which combine increased selectivity with higher permeation rates.   Less effort has been devoted to studying membrane separation of gases as a system.   In commenting on this subject I am conscious of a total lack of experience, but this fact may help me to be unbiased.

Most modern membranes do not depend for their separation function on differential rates of gaseous diffusion through a porous matrix.   Rather they use a porous film to support a very thin and essentially non-porous coating.   The coating is chosen for its solvent properties with respect to the permeating gas, so that the permeate dissolves in it on one face, diffuses through, and evaporates on the other face.   Hence membrane separation resembles absorption separation (already discussed) in which the solvent does not circulate but exists as a stationary film.   The feed gas mixture is exposed to one side of the film, and regeneration by pressure reduction occurs continuously on the other side.   The coating should be as thin as possible since thickness contributes little to the separation but accounts for most of the pressure drop (energy consumption).

Asymmetric membranes have been developed on this basis but initially from the same material which is formed as a porous structure over most of the membrane thickness with a dense layer at the surface.   But the primary duty of the main membrane is to support the coating against the process pressure, whilst being as porous as possible.   Hence it needs to have an anisotropic honeycomb structure with pores extending through the membrane, but with mechanical strength to carry the gas pressure load (whether as a hollow fibre or as a flat sheet bridging between spacers).   For

this purpose the membrane thickness will probably require to be
between 0.01 and 0.1 mm.

By comparison the coating of non-porous absorbent will have a
thickness of only 0.01 to 0.1 μm.   The difference of thickness of
about three order of magnitude between the membrane and coating is
probably too great to be bridged directly, so an ideal membrane
will need an intermediate support structure to carry the coating.

System considerations can profitably commence by readdressing
Fig. 1, the model of ideal gas separation.   It requires two ideal
membranes, one asymmetric membrane in which component $\underline{a}$ is highly
soluble and $\underline{b}$ virtually insoluble, and a second asymmetric membrane
with the converse properties.   Even if these membranes existed
there is a further condition to be fulfilled for ideal separation –
that the pistons move at an infinitely slow rate.   The reason is
not simply the pressure drop across the membrane but is the
accumulation of the non-permeating molecules at the membrane
surface and the speed with which they can back-diffuse.   This
accumulation will reduce the partial pressure of the permeate and
so will reduce the permeation rate until a diffusional balance has
been achieved.   An analogous situation occurs with the
condensation of a vapour containing a small amount of inert gas.
One per cent of inert gas can halve the condensation rate.   This
gas phase resistance must be present in all membrane separations
but seems to be largely ignored.   Of course if permeability is
very low the effect will be negligible; high permeabilities will
magnify it.

Referring back to Fig. 1, the situation could be helped by
mounting a propeller above the membranes, but even so the thickness
of the boundary layer above the membrane surfaces can never be
reduced to zero.

Membrane units normally comprise either spirally wound flat
sheets separated by spacers, or bundles of hollow fibres.
Discussion will be restricted to the latter, but the principles are
the same.   The feed gas flows over the outside of the hollow
fibres (typically 0.5 mm diameter and about 3 m. long) within the
enclosing pressure vessel.   The separation characteristics of such
a unit will be discussed by considering the example (10) of natural
gas sweetening and by simplifying the unit to a single fibre (Fig.
10).

At the inlet end the nominal driving force for CO_2 diffusion
is 3.6 bar and the permeation rate will be high.   In practice the
accumulation of methane against the fibre coating will be
considerable and the resulting CO_2 driving force will be reduced.
This diffusion barrier will persist along the fibre and
demonstrates the need for a spacer between the fibres which will
not only define the flow path for the residual gas, but will help

to mix it locally.

The residue gas should flow from one end of the cell to the other with minimum axial mixing so that its purity (with respect to the non-permeating gases) is as high as possible, and so that the driving force for permeation is maximised. A snag with membrane separation is the limited purity of the residual gas. The permeating component must always have a substantially larger partial pressure in the residue than in the permeate - the reverse of what one would wish.

In the example the reduction of CO_2 content to 2% is possible only because there is a pressure difference of about 57 bar between residue and permeate and because the permeate contains only 63.4%CO_2. If the membrane was improved so that the permeate contained 75% CO_2 then the permeate would require to be withdrawn under partial vacuum. In practice the CO_2 content of the permeate can be upgraded (or the loss of CH_4 reduced) by recompressing the permeate to somewhat over 60 bar and feeding it to a second membrane cell. The permeate from this second cell can then contain more than 80% CO_2 whilst the residue has about the same composition (and pressure) as the original feed, and so can be recycled into it.

The amount of permeate from the first cell to be compressed is about 15% of the original feed, but the work required is considerable due to the large compression ratio. In fact when these processes are assessed in terms of the minimum work of separation (Fig. 2) it is apparent that their thermodynamic efficiency is very low.

The attraction of membrane separation, as demonstrated by this example, is that it can achieve an enrichment ratio between permeate and residue of 36.6/2.0=18.3, and a methane recovery of 90.2% in a single stage. The high enrichment is gained at the expense of low thermodynamic efficiency, and this inefficiency is multiplied if units are staged.

One may conclude that membranes are attractive for gas separation when:

(i) A membrane is available with a uniquely high permeability for the component to be separated.

(ii) Only a rough separation is needed.

(iii) The feed gas is at an elevated pressure.

There are cases in which membrane separations are used profitably in combination with another separation means.

8 CONCLUSIONS

The final word on industrial gas separation has not yet been spoken. Wholesome competition exists between continuous and intermittent processes, and between low temperatures and high. Much good engineering is evident in present achievements, but there is scope for more.

It is to the credit of the industry that it has generated most of its own developments. However there has also been a good partnership with academe, and the portents are good for its continuance.

REFERENCES
1. B.Linnhoff, Chem.Eng.Res.Des., 1983, 61 207.

2. K.B.Wilson, D.W.Woodward and D.C.Erickson, Proc. 12th Int. Cryogenic Engng. Conf., Southampton, July 1988, (published Butterworths), p.355.

3. G.G.Haselden, Trans.Instn.Chem.Engrs., 1958, 36, 123.

4. Q.M.Siddique, Chemical Engineer, 1985 (June), No.415, 26.

5. B.R.Dunbobbin and W.R.Brown, Gas Sepn. and Purification, 1987, 1, 23.

6. R.T.Yang, 'Gas Separation by Adsorption Processes', Butterworth, U.S.A., 1986.

7. C.Y.Pan and D.Basmadjian, Chem.Eng.Sci., 1967, 22, 285.

8. D.H.White and P.G.Barkley, Chem.Eng.Prog., 1989, 85, No.1, 25.

9. K.J.Doshi, C.H.Katira and H.A.Stewart, AIChE Symposium Series, 19, 67, No.117, 90.

10. R.W.Spillman, Chem.Eng.Prog., 1989, 85, No.1., 41.

Gas Separation by Distillation

M. J. Lockett

UNION CARBIDE INDUSTRIAL GASES, INC., TONAWANDA, NY 14151, USA

Abstract

Some recent, and some not so recent, developments are
reviewed relating to the use of distillation for the
separation of gases. Topics covered include C3 and C2
splitters, demethanizers, nitrogen rejection units, air
separation columns and rotating devices. The emphasis is
primarily on the different types of column internals,
such as trays and packing, which are used to make the
separation.

1 INTRODUCTION

Joseph Priestley left England for America in 1794 after a
Birmingham mob burned his house, church and laboratory.
It is not necessary to go to quite those lengths nowadays
to encourage people to make the same move and, as one who
has followed in Priestley's footsteps, it gives me
particular pleasure to be asked to speak at this
conference held in his honor.

In preparing this paper on gas separation by distillation it was first necessary to decide on which of the
light hydrocarbons should be included. Ruhemann (1)
draws the line at butane, including it and more volatile
hydrocarbons in the definition of a gaseous mixture.
However, the subject of C4 separations is extensive and
merits a complete paper in its own right, and so the
present paper deals only with C3's and lighter hydrocarbons together with the separation of the major
constituents of air. To avoid an overly superficial
review of too many topics, the emphasis of the paper is

on hardware and column internals since these are of particular interest to the author. No attempt is made to cover other important topics such as new process cycles, column control or boiling and condensing.

2   PROPYLENE - PROPANE AND ETHYLENE - ETHANE SPLITTERS

The last two or three years have witnessed a soaring demand for both ethylene and propylene with plants world-wide being operated at close to their nameplate capacity. Although some new plants have been announced, producers have generally been wary of investing in new capacity having learned hard lessons from the overcapacity situation which prevailed in the early 1980's. Instead there has been a remarkable surge of activity focused on debottlenecking existing equipment and often the distillation columns have proved to be the key items which have limited plant capacity.

Although each column revamping job is different with its own unique problems, four different strategies can be identified for revamping C3 and C2 splitters for increased capacity.

a)   When a distillation column contains trays, it is often possible to increase capacity simply by replacing the existing trays with trays of the same type but which have been better optimized for higher capacity. For example, the weir heights can be reduced. The bubbling and downcomer areas can be reallocated. Or, for sieve trays, the open hole area can be increased to relieve downcomer backup at the expense of providing less turndown. Each of the above can directionally increase column capacity when judiciously implemented. The author has experience of a propylene-propane splitter in which the original 4-pass sieve trays were modified twice, each time to provide an incremental increase in column capacity.

b)   Frequently it is possible to operate a distillation column at a more favorable point on the theoretical trays versus reflux ratio curve to achieve a capacity increase. For fixed product and feed specifications typical curves are shown in Figure 1.

Ethylene-ethane columns tend to be operated at reflux ratios close to the minimum because reflux and boilup are provided via a refrigeration loop which is expensive.

Conversely, for propylene-propane columns located in ethylene plants, the reboiler is heated by quench water which is virtually free and condensation is provided by cooling water. Consequently for this separation, the economic optimum reflux ratio tends to be considerably greater than the minimum value.

Consider now a propylene-propane splitter operating at a particular point on its theoretical trays versus reflux ratio curve. Suppose conceptually that the column were to be retrayed with the same type of trays but at a reduced tray spacing. The number of theoretical trays will increase, the required reflux ratio will fall and the internal vapor and liquid loads within the column will be reduced. Although trays at a reduced tray spacing can handle lower loads without flooding, often the increase in capacity because of the reduction in reflux ratio is greater than the reduction in capacity caused by the reduced tray spacing. The net effect can be an increase in the column capacity such that an increased feed rate can be accommodated. There is clearly an optimum tray spacing which gives the maximum column capacity for given feed and product specifications. Note that this strategy for increasing column capacity is not dependent on the availability of new improved types of column internals. It can be implemented with almost any type of column internals.

For example, a similar approach can be applied for revamping columns containing packing. For random packing it is the ring size and for structured packing the packing density which is optimized in place of the tray spacing. Thus a decrease in ring size or an increase in packing density will reduce the height of an equivalent theoretical plate (HETP) at the expense of a reduction in the capacity of the packing. As for trays, the overall net effect can be to increase column capacity if the required reflux ratio falls sufficiently. At first sight, since packing is fabricated in a limited number of discrete sizes, whereas tray spacing can be continuously varied, the possibility of optimizing capacity might appear to be more limited for packing than for trays. In practice, it is often desirable to reuse as many tray support rings as possible when revamping columns containing trays. This minimizes the cost and downtime. Thus the new trays are often installed at half or perhaps two-thirds the original spacing. Under this constraint there is a close analogy between the revamping of trayed and packed columns using this strategy.

c) Column revamping for increased capacity can be
accomplished by replacing a given type of column
internals with a similar but improved version. For
example, one of the modern types of random packing, such
as Interlox Metal Tower Packing or Cascade Minirings, can
be used to replace Pall rings (3). Multiple Downcomer
trays or perhaps Perform-Kontact trays can be used to
replace 2 or 4 pass sieve or valve trays (4). It sometimes turns out that the replacement column internals do
have a higher capacity but at the expense of a lower
separation performance. This can often be overcome by
increasing the reflux ratio and so operating at a
different point on the theoretical trays versus reflux
ratio curve, particularly if the original design is close
to minimum reflux. Note that strategy (2), in which an
increase of theoretical trays is achieved, works best for
columns operating well away from minimum reflux, such as
C3 splitters. This is because, as shown in Figure 1, the
slope of the C3 curve is such that a large change in
reflux results from a small change in theoretical trays.
Strategy (3), in which a reduction in theoretical trays
may occur when higher capacity internals are substituted,
works best when the design is close to minimum reflux
such as in C2 splitters. In this case as shown in Figure
1, a shortfall in theoretical trays can be compensated by
only a small change in reflux. A recent paper deals in
more detail with this approach for revamping ethylene-
ethane columns (2). Note also that strategies (2) and
(3) are quite often implemented together.

d) Finally we have the most radical and dramatic type of
column revamping which involves replacement of one type
of column internals with a completely different type, for
example replacement of trays by packing or vice-versa.
This type of column revamp tends to be extensively
reported in vendor's literature. Very often, strategy
(2) above is implemented at the same time, making it
unclear just how much of the reported capacity increase
can be attributed to the inherently superior capacity of
the new type of column internals.

In fact for high pressure columns used for gas
separation there appears to be only a marginal difference
between the capacity of well designed trays and properly
installed and functioning packing when evaluated at the
same separation efficiency. Pilot plant data have been
reported on the performance of structured packing when
used for the distillation of light hydrocarbons (5).
However, there have been recent reports that in industrial columns structured packing appears to experience a

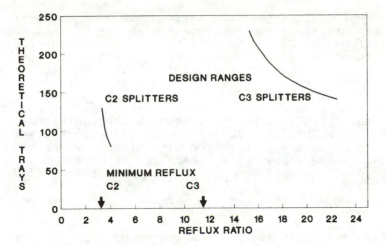

**FIG.1 THEORETICAL TRAYS VS. REFLUX RATIO
C2 AND C3 SPLITTERS**

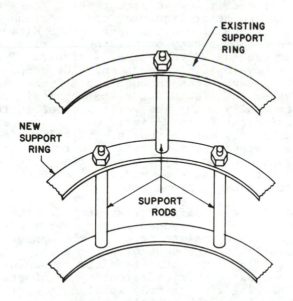

FIG.2 TRAY SUPPORT RING SYSTEM

sharp decline in separating performance when operated at
the high pressures and consequent high liquid loads such
as are found in C3 and C2 splitters. The reasons for
this are not completely clear and the subject is currently undergoing intense investigation. Possibly it arises
from vapor backmixing caused by the high liquid loads
which has been shown to have an adverse effect on the
separation performance of random packing (6). Although
there have been a number of instances where disappoint-
ingly high HETP's were obtained when trayed propylene and
ethylene columns were revamped with structured packing,
similar problems have so far not been reported when
random packings have been used in these services (7).

3   DEMETHANIZERS

Demethanizers frequently are a source of capacity limitation in ethylene plants and revamping them to increase
capacity can present a formidable challenge. They
operate at about 70% of the critical pressure and at such
a high pressure the distinction between vapor and liquid
tends to become blurred. At the top of the column the
surface tension can be as low as 1 to 2 dynes/cm and
difficult to predict. Low surface tension tends to
encourage entrainment of liquid by the vapor and entrainment of vapor by the liquid. Under such conditions
accurate prediction of tray or packing capacity, tray
efficiency or packing HETP becomes difficult. Typically
demethanizers have multiple feeds and wide composition
and temperature variations through the column. As a
result the vapor and liquid loads and physical properties
differ substantially among the various zones in the
column. Prediction of vapor-liquid equilibrium and
physical properties, such as phase enthalpies, is not
straightforward and differences exist between the prediction methods used by the major engineering contractors,
the operating companies, the vendors of computer process
design packages and equipment vendors.

In spite of these difficulties, many demethanizers
have been successfully revamped in the last few years.
Revamping generally consists of using the appropriate
type of internals in each zone of the column. Thus it is
not unusual to find demethanizers after revamping
containing combinations of different types of trays and
packings. The ring size or tray spacing is optimized in
each zone to maximize the overall column capacity. Since
demethanizers operate at low temperatures and high
pressures, welding in new tray support rings at a new

optimized tray spacing requires post-weld heat treatment
of the vessel. This tends to be prohibitively expensive
and so alternative methods have been devised to retray
such columns at a different tray spacing without the
necessity of welding. One such method is shown in
Figure 2 where the existing tray rings are used to
support new rings using a system of support rods which
are bolted to the existing rings. Although this makes
the revamping operation more complicated than usual, the
results have been found to be worthwhile, since in this
way the new trays can be installed at an optimum spacing
and column capacity can thereby be maximized.

4   NITROGEN REJECTION UNITS

A number of projects have been and are being implemented,
particularly in the USA, in which nitrogen is injected
into an oil reservoir to enhance the recovery of the
oil. The produced gas leaving the reservoir contains
methane, some heavier hydrocarbons, and a variable amount
of nitrogen. It can be economically attractive to
recover the nitrogen for subsequent reinjection into the
reservoir, and the heavier hydrocarbons are recovered as
liquids also. The separation of nitrogen and liquids
from the methane is accomplished by cryogenic distillation in a nitrogen rejection unit (NRU). The design of
the distillation columns in the NRU presents an interesting challenge because the columns have to be capable of
operating efficiently over an extremely wide range of
operating conditions. This is mainly because at the
beginning of the project life there is only about 8% of
nitrogen in the feed to the NRU but in subsequent years
this can rise to perhaps 80%. As a consequence, vapor
and liquid flowrates in the distillation columns can vary
by a factor of 20 to 1. The preferred type of column
internals where very high rangeability is needed are
bubble cap trays because weeping can be entirely
eliminated from such trays, and a number of patents have
been awarded recently for improved types of bubble cap
trays for NRU applications.

   Bennett et al described a bubble cap tray having a
row of special bubble caps adjacent to the inlet weir
(8). These caps have a slot opening under the skirt of
the cap as normal, and in addition there are a series of
small holes partway down the cap. At very low vapor
flowrates bubbling takes place only from the holes in the
cap. In this way sufficient vapor-liquid contact can be
provided for adequate mass transfer even at the lowest

vapor and liquid flowrates encountered. Without these
special bubble caps at the liquid inlet, vapor bubbling
can be non-uniform at low flowrates allowing the possibility of liquid and vapor bypassing with an attendant
low tray efficiency.

Another type of bubble cap tray for NRU applications
has been developed consisting of bubble caps having slots
whose elevation off the tray floor is matched to the
liquid hydraulic gradient at full-load flowrates (9).
Again a special row of caps is used at the liquid inlet
which have extra-high slots such that the liquid seal
over these slots is less than for the remainder of the
caps. At extreme vapor turndown conditions the inlet
caps continue to bubble uniformly and efficient vapor-
liquid contacting is still maintained. Additional
features of the caps are the use of triangular slots and
the provision of equal open area for each cap. The
latter tends to ensure uniform vapor flow through the
caps and satisfactory tray efficiency at full-load
flowrates.

Presently the turndown requirement in some NRU
applications is larger than can be provided by packed
columns. However, the turndown limit of packed columns
in these applications is not really very well defined.
It would be interesting to know what this limit really
is, with adequate attention given to wetting of the
packing by liquid, perhaps with an appropriate surface
texture on the packing, and with the provision of special
high-turndown distributors.

5 AIR SEPARATION BY DISTILLATION

Notwithstanding some recent advances in membranes and
pressure swing adsorption, air separation by distillation
continues to be the most cost-effective route for medium
to large tonnage applications and is even preferred on
the small scale when high purity products are required.
This situation is expected to continue into the indef-
inite future. Furthermore, significant advances have
been and are being made in distillation technology for
air separation which tend to get overlooked in comparison
with the attention currently being given to membranes and
PSA.

Because of the recent decline in the cost of structured packing, its use for air separation has now become
viable. The concept of using structured packing for air

separation is not new. Just after the second World War,
Weedman and Dodge (10) and also Aston at al (11) provided
extensive data on the performance of a wide range of
packings, including both random and structured packings,
when separating air with column diameters in the range of
2-12 inches. The HETP's they reported were as low as 1.8
inches. More recently Yamanishi and Kinoshita (12)
reported HETP's of 5.5 cm using Dixon rings for N_2-Ar
separation.

The use of packing for the distillation of air
instead of trays provides a substantial reduction in the
power required to drive the air compressor. A typical
pressure drop summary for a trayed air separation plant
is shown in Figure 3. Packing has a pressure drop which
is about 20% that of trays. Thus replacing trays by
packing cuts the upper column pressure drop from 6.4 to
about 1.1 psi and the lower column pressure drop from 2.7
to about 0.7 psi. Because of the different vapor pressure-
temperature relationships of oxygen and nitrogen, the
reduction in upper column pressure drop of 5.3 psi is
magnified by a factor of about 3 across the reboiler-
condenser which links the upper and lower columns. The
net result is that packing both the upper and lower
columns can reduce the compressor discharge pressure from
93 to about 75 psia. The power saving for the air
compressor is thus about 12.5%. For a plant producing
500 tpd of oxygen the electrical power savings are
approximately $250,000 per annum evaluated at 5c/kwh.
Additionally, because of the lower absolute pressures
which are obtained in both the upper and lower columns,
the vapor-liquid equilibrium relationships become more
favorable and the recovery of argon is increased.

In spite of its recent cost reduction, structured
packing still costs more than trays. The additional cost
of packing over trays has to be offset against the power
savings and additional argon recovery. Although random
packing is considerably less costly than structured packing, its performance is less reliable for large scale
demanding separations such as the distillation of air.
Thus the use of structured packing is preferred and there
is a great incentive to gain as much of the power savings
as possible but at the same time to minimize the amount
of packing which is used.

It is found that if packing is used throughout the
upper column while retaining trays in the lower column
and also in the low ratio (or crude argon) column, the
bulk of the air compressor power savings are still obtained

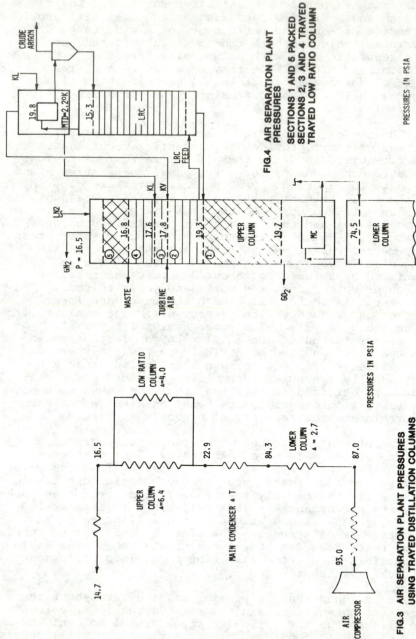

FIG.4 AIR SEPARATION PLANT PRESSURES

SECTIONS 1 AND 5 PACKED
SECTIONS 2, 3 AND 4 TRAYED
TRAYED LOW RATIO COLUMN

PRESSURES IN PSIA

FIG.3 AIR SEPARATION PLANT PRESSURES
USING TRAYED DISTILLATION COLUMNS

PRESSURES IN PSIA

while the amount of packing used is considerably reduced. Unfortunately in this case plant operational
problems are introduced because the pressure at the top
of the low ratio column falls below atmospheric. This is
undesirable, not least because of the possibility of
argon product contamination. Furthermore, the available
temperature difference across the low ratio condenser
heat transfer surface falls to approximately 0.7°K, which
is too low for effective operation.

One solution is to use packing in all or part of the
low ratio column, but this involves the use of additional
expensive packing without any direct benefit for power
saving (13). An alternative solution which has also been
proposed is to use packing and trays in combination in
the upper column while retaining trays in the low ratio
column. One example of this is shown in Figure 4. In
this scheme plant operational problems are avoided at the
top of the low ratio column and at the same time substantial power savings are still obtained at the air compressor.

6 AIR SEPARATION IN ROTATING EQUIPMENT

Several concepts for single and two stage to orbit aerospace planes operate on air-breathing propulsion initially
and on rocket propulsion for the remainder of their flight
into orbit. During the air-breathing stage, air will be
collected from the atmosphere, liquefied and stored onboard
for later use as an oxidizer in the rocket engines. A
development of this idea requires additional onboard
enrichment of the air to produce liquid oxygen. During the
1960's there was a great deal of classified work carried
out in the USA on the aerospace plane program under the
sponsorship of the U.S. Air Force, and most of it was
declassified in the 1970's.

The concept of air separation in a rotating double
column was initially suggested by workers at General
Dynamics in 1960 although the patent did not appear until
1973 (14). Throughout most of the 1960's, studies were
performed by a number of organizations and the Linde
Division of Union Carbide was contracted to develop, build
and test a large scale device to separate air in rotating
columns. Although this work has been part of the public
record for more than 15 years, it has not been widely
reported and so it is worthwhile mentioning it briefly
here. More details are available in the original reports
(15).

The basic air collection and enrichment system (ACES) is shown in Figure 5. Both the high and low pressure columns, the reboiler-condenser and the reflux condenser were rotated on a common hollow shaft which was also used as a conduit for the fluids. The process cycle resembled the standard double column air separation process cycle except that hydrogen was used as the source of refrigeration in the reflux condenser. The columns contained sieve trays at spacings of about 20-40mm and gravitational forces up to 350 times that of normal gravity were used. An illustration of the sieve trays is shown in Figure 6(16). The reboiler-condenser also operated in the high gravitational field. It consisted of a multiplicity of 7mm id tubes having a porous coating on both internal and external surfaces to enhance boiling and condensation. A diagramatic representation of the reboiler-condenser is shown in Figure 7.

A very considerable amount of development work was undertaken by Union Carbide which in 1965 culminated in the construction and successful operation of the boiler-plate rotor shown in Figure 8. It was demonstrated that the volume of the distillation equipment could be reduced by a factor of 20 to 40 by the use of high gravitational fields. Recently there has been a revival of interest in the aerospace plane and work is continuing at Union Carbide on further developing and perfecting the concept of air separation in rotating equipment.

Conclusion

The purpose of this paper has been to demonstrate that new innovations and challenges are still to be found in the technology of gas separation by distillation. Ongoing developments are making distillation even more cost effective and reinforcing its position as the preeminent technology for gas separation.

Acknowledgement

Special thanks are due to Mr. C. F. Gottzmann who helped the author become familiar with Union Carbide's past work on rotary air separation.

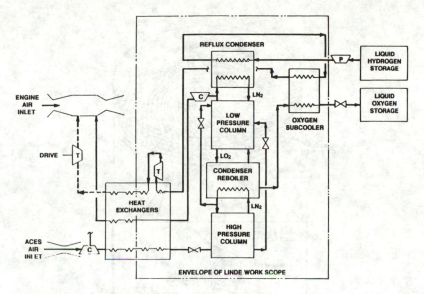

FIG.5 DOUBLE COLUMN DISTILLATION - ACES

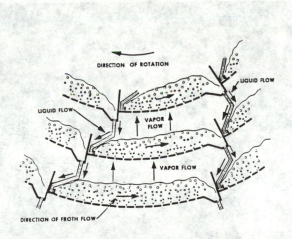

FIG.6 ROTATING DISTILLATION TRAYS

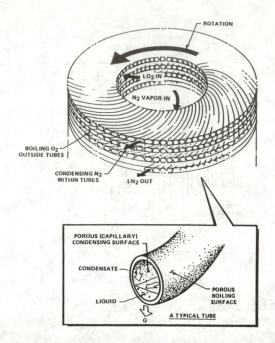

FIG.7 ENHANCED ROTARY REBOILER-CONDENSER

FIG.8 BOILERPLATE ROTOR

REFERENCES

1. Ruhemann, M. (1949). The Separation of Gases, Oxford University Press.

2. O'Neill, P. S., Lockett, M. J. and Navarre, J.L. (1988). Experience With High Performance Process Equipment In Large Ethylene Plants, Proc. Symposium on Large Chemical Plants, 6-7 Oct., Bruges, Belgium, Royal Flemish Society of Engineers.

3. Strigle, R. F. (1987). Random Packings and Packed Towers, Gulf, Houston.

4. Resetarits, M. R., Agnello, J., Lockett, M. J. and Kirkpatrick, H. L. (1988). Retraying Increases C_3 Splitter Column Capacity, Oil & Gas J., June 6, 54.

5. Spiegel, L. and Bomio, P. (1987). Hochdruck - Rektifikation und - Absorption Mit Der Geordneten Packung Mellapak, Chem. Ing. Tech., 59, No. 2, 130.

6. Woodburn, E. T. (1974). Gas Phase Axial Mixing at Extremely High Irrigation Rates in a Large Packed Absorption Tower, AIChE J., 20, No. 5, 1003.

7. Dolan, M. J. (1984). Advances in Ethylene Processing via IMTP Technology, AIChE Meeting, November 30, San Francisco.

8. Bennett, D. L., Edwards, T. J. and Rowles, H. C. (1985). Perforated Bubble Caps for Enhanced Vapor-Liquid Contact on a Distillation Tray, US Patent 4,510,023.

9. Lockett, M. J., Summers, D. R., Smith, V. C. and Upchurch, J. C. (1987). High Turndown Bubble Cap Tray, US Patent 4,711,745.

10. Weedman, J. A. and Dodge, B. F. (1947). Rectification of Liquid Air in a Packed Column, Ind. & Eng. Chem., 39, No. 6, 732.

11. Aston, J. G., Lobo, W. E. and Williams, B. (1947). Liquid Air Fractionation, Ind. & Eng. Chem., 39, No. 6, 718.

12. Yamanishi, T. and Kinoshita, M. (1984). Preliminary Experimental Study for Cryogenic Distillation Column with Small Inner Diameter, J. Nuc. Sci. & Tech., 21, No. 1, 61.

13. Victor, R. A. and Lockett, M. J. (1989). Double Column Air Separation Process with Hybrid Upper Column, US Patent 4,838,913.

14. Nau, R. A. and Campbell, S. A. (1973). Rotary Separator, US Patent 3,779,452.

15. Gottzmann, C. F. et. al. (1966). Air Separator Test Program, U.S. Air Force Air Propulsion Laboratory Report AFAPL-TR66-92.

16. Bonnet, F. W. (1974). Rotary Fluid Contactor, US Patent 3,809,375.

Advances in Cryogenic Air Separation

T. D. Atkinson

BOC LIMITED, 10 PRIESTLEY ROAD, GUILDFORD GU2 5XY, UK

T. Rathbone

BOC CRYOPLANTS, 30 PRIESTLEY ROAD, GUILDFORD GU2 5YH, UK

ABSTRACT

Considerable growth in the markets for oxygen and
nitrogen has been achieved by increasing the efficiency
of cryogenic production plants, and hence making
available lower cost product. Over the past 10 years
the efficiency of air separation has been increased by
25%.

The factors which have contributed to these
improvements include the availability of modern computer
techniques and data systems, renewed interest in 2nd Law
methods as demonstrated by the work of Linnhoff et
al[1], and the availability of more efficient cryogenic
equipment and compressors. Particular attention has
been applied to pushing the distillation process to the
limits of efficiency, and ensuring that the column is
integrated effectively into the refrigeration cycle.

An extension of this work has been to examine the
potential for utilising work which can be recovered when
streams of unequal composition are mixed. This option
can be integrated into separation cycles to give a wider
range of possible products for a given operating cost.

Another application for fundamental thermodynamic
analysis is to identify opportunities for reducing the
cost of air-separation by integrating the cryogenic
cycle with other production cycles, and work in this
area paves the way for many exciting advances in air
separation.

1 INTRODUCTION

Improving the efficiency of air-separation plants is an
important objective for industrial gas companies.
Reduced power consumption enables lower production
costs, and the resultant lower cost product can be used
to stimulate market growth.

Efficiency improvements can be achieved either
through developing more advanced equipment, or by
finding ways to optimise the design of the process
cycle. In this paper we describe some recent work which
sought to examine the losses inherent in existing
process cycles, so that more efficient designs could be
developed. The guiding light for this work was the
second law of thermodynamics, which has been applied by
Linnhoff and many other workers to develop methods for
reducing losses in a wide variety of process cycles.

Our paper starts with an analysis of losses in a
cryogenic air separation cycle. The method we developed
to measure these losses inspired us to take a fresh look
at the problem of producing argon at low cost, and our
solution is described later. We also show how close
analysis of losses in distillation enable innovations to
existing process cycles to be made with consequent power
savings. Finally we discuss how further improvements
are possible by considering an air-separation plant in
the context of the processes to which it delivers
product. For example, an air-separation plant
integrated with a coal gasification system can be more
efficient than a plant in which both units are optimised
independently.

2 CAUSES OF LOST WORK IN A CRYOGENIC CYCLE

In an 'ideal' process, all changes are achieved
reversibly. For example if a process stream were
compressed from P1 to P2 the work required could be
totally recovered if the stream were expanded back from
P2 to P1. The reversible process also defines the
"minimum" work required to achieve a change. This is
because if it were possible to recover more work from
the expansion than the work required for compression
then energy would be created from nothing, and this
violates the second law of thermodynamics.

If more work than the minimum required by an ideal
system is used to compress a stream from P1 to P2, then
even if the expansion is achieved reversibly, this will

not fully recover the work of compression.   A
thermodynamic loss has been incurred.

Thermodynamic losses occur throughout an air
separation plant. By understanding the nature of these
losses, they can be minimised, hence increasing the
overall efficiency of the cycle. Losses can be divided
into the three categories described below:-

a)   Irreversible expansion or compression

When a gas flows through a pipe there will be some
pressure drop due to friction. Although the
pressure falls, there is no recovery of energy, so
work is lost. Similar losses also occur in
machinery such as compressors and expanders.

b)   Irreversible heat transfer

In heat exchangers, heat flows from a warm stream
to a colder stream. Since work would be required
to transfer the heat back up to the warm stream,
this heat transfer represents a loss. Similar
losses also occur in distillation, where warm
streams mix with colder streams on a tray.

c)   Irreversible Mass transfer

In many air-separation plants, a waste nitrogen
stream is produced, and vented to atmosphere. The
mixing of the nitrogen with the air is a loss,
since work would be required to recover the
nitrogen stream from the air. In addition to
process mixers such as this, losses also occur when
streams which are not in thermodynamic equilibrium
are mixed on distillation trays.

3 ANALYSIS OF LOST WORK

Thermodynamic losses can be measured by comparing the
work actually required to achieve a process change, to
the minimum theoretical work required to produce that
change. Since lost work can be divided into three
components it is also useful to divide process work into
similar components.

This is achieved by considering three reversible
machines, which can change pressure, temperature, or
composition. This system has been described in detail
by Atkinson[2].

a)   Pressure
       A reversible compressor is a well know model,
       against which to evaluate the thermodynamic cost of
       actual pressure changes. The minimum work required
       to compress a fluid from P2 to P1 is :

$$Work = RTo \ \ln \frac{(P2)}{(P1)}$$

b)   Heat Pump
       A reversible heat pump transfers heat Q from a
       temperature T to To, using :

$$Work = \frac{Q(To - T)}{T}$$

c)   Separation/Mixing
       A reversible separation device utilises the
       perfectly selective properties of an ideal
       membrane, to separate a gas into its components.
       Compressors return these components to the pressure
       of the feed.

       The work required is given by the expression:

$$Work = RTo \sum xi \ln xi$$

       A conceptual approach to assessing lost work in a
cycle is to calculate the amount of reversible work
needed to return the products to the conditions of the
feed. In the simple example of pressure drop in a pipe,
the calculation would show how much work a reversible
compressor would need to overcome the pressure drop.
The following example shows how the results appear for
an air-separation plant, and illustrates how these
results can be used.

Example:
Three Component Analysis Of Air-Separation Plant

       A simplified diagram of a conventional
air-separation plant is show in Figure 1. Compressed
air is cooled in the heat exchanger and fed to the
high pressure distillation column. A liquid nitrogen
stream, and an oxygen rich liquid stream are
withdrawn from the high pressure column, and expanded
through Joule-Thompson valves into a low pressure
column. Oxygen and nitrogen product streams are
taken from this column, and returned to ambient
temperature in the heat exchanger. An argon rich

stream is withdrawn from the column and purified in a
side column. In addition a turbine (not shown on
Figure 1) must be provided in order to supply the
refrigeration needs of the system. Either air or
nitrogen may be used as working fluid in this machine.

An analysis of major components of this cycle is
shown in Table 1. The first line of the table shows
results for the heat exchangers, the second shows the
two Joule-Thompson valves, and the third shows the
three distillation columns.

TABLE 1
Derivation of lost work from "Three Component" Analysis

	Pressure Work MW	Heat Pump Work MW	Separation Work MW	Lost Work MW
Heat Exchangers	-0.40	-1.05		1.45
J/T Valves	-7.0	6.38		0.62
Column System	0.12	-4.66	2.10	2.44

The first column of Table 1 shows the input or
output calculated for a reversible heat pump which
returns all the products back to their feed condition.
The second column shows similar results for a reversible
heat pump, and the third shows results for a reversible
composition change. The total lost work is calculated
by adding inputs or outputs for the three work
components.

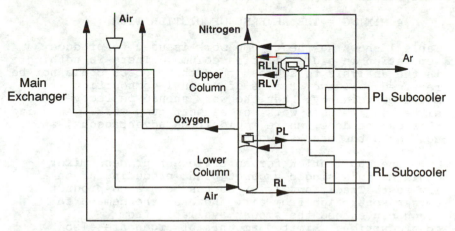

Figure 1 Simplified Air Separation Cycle

The results for the heat exchanger show that
1.45 MW of work have been lost, and that 0.4 MW is due
to pressure drops, the rest being due to temperature
differences. The results for the Joule-Thompson valves
show that a reversible compressor operating at room
temperature would need to provide 7.0 MW to overcome the
pressure drop in the valve. However because the fluid
has cooled in the valve, this is worth 6.38 MW of work
from a reversible heat pump. Work has been effectively
transferred from being available as pressure, to being
available as a heat pump potential. The losses incurred
in this transformation are only 0.62 MW, less than 10%
of the input.

In the distillation column a further transformation
is observed. A heat pump potential of 4.66 MW is used
to separate the fluids, and a reversible separation work
of 2.1 MW would have been required to produce this
separation. Whereas vapour streams lose pressure in a
column, liquid streams increase in pressure, and on
balance a small recovery of pressure work is also
observed in the columns.

The value of this type of analysis is that it
provides insight into both the source of lost work, and
also the processes as which incur the lost work. The
example which follows shows how this approach enabled us
to develop cycles based on efficient process mixing.
However, there are many other ways in which this
analysis can help to improve efficiency.

4 MIXING – THE REVERSE OF DISTILLATION

Table 1 shows how heat pump work is used to produce
separation in a distillation column. There is nothing
in the analysis to suggest that the process could not be
reversed. If two streams of unequal composition are
mixed efficiently, then the work should be recoverable,
either as pressure work, or as heat pump work. Assuming
this can be done, where would it be advantageous, and
how would the process work?

One application for an efficient process mixer
would be in a single product plant producing argon at
low cost. Traditionally argon has been a co-product of
oxygen separation from air. However the demand for
argon can exceed the amount available from
co-production. Limited amounts of argon are also
recovered from ammonia production plants; however it is
sometimes necessary for air separation plants to be run

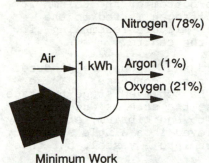

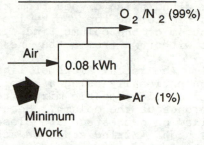

<u>Figure 2</u> Methods of Argon Production

to produce only the argon, with both oxygen and nitrogen venting back to atmosphere.

   Figure 2 shows that this is a very inefficient way to produce argon. The theoretical minimum work to separate air completely is over ten times greater than that required to merely remove argon from the air.

   Using conventional distillation, it is not possible to design a plant which can produce a high yield of argon without requiring a total separation of air into oxygen and nitrogen. This is because argon has a volatility which lies between oxygen and nitrogen. Therefore in any distillation sequence if the purity of the nitrogen product were reduced, the first impurity it would contain would be argon, and virtually all the argon would be lost in the nitrogen before a significant amount of oxygen was also present in the stream.

   If it is accepted that oxygen and nitrogen must be separated to produce argon, one way to reduce overall power consumption would be to re-mix the oxygen and nitrogen. If this mixing is efficient then the resultant work can provide some of the work for the separation process.

<u>Efficient Mixing</u>

   Figure 3 shows a device which can achieve efficient mixing. It is similar to a cryogenic distillation

column, except that many features are reversed. Instead
of withdrawing oxygen from the bottom of the column it
is fed to the top. Likewise nitrogen is fed to the
bottom of the column, and a gas mixture resembling air
is withdrawn from a tray near the centre.

The reflux flows of liquid and vapour are produced
conventionally using a reboiler and condenser. These
reflux flows enable mixing to be achieved stage by stage
throughout the column. Because the streams which are
mixed on each stage are close to equilibrium, efficiency
is high. Mixing could be achieved in a single stage, by
introducing oxygen and nitrogen into a vessel, but this
would be very inefficient.

An important feature of the mixing column is that
the condenser is warmer than the reboiler, since at a
given pressure, saturated liquid oxygen is warmer than
liquid nitrogen. Heat which is transferred from the
reboiler to the condenser is upgraded to a higher
temperature, and this heat pumping duty is the output
from the mixing process.

An Argon Generator Cycle

Figure 4 shows one cycle which utilises efficient
mixing to produce argon. The production cost of argon
is significantly less than a conventional air separation
plant operated solely to produce argon.

At the centre of the cycle, a distillation column

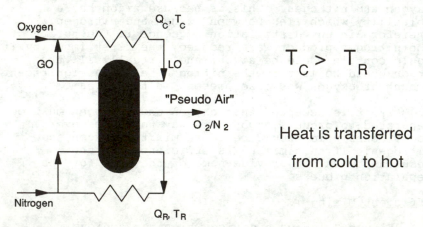

$$T_C > T_R$$

Heat is transferred
from cold to hot

Figure 3 The Mixing Column – A Distillation Column in Reverse

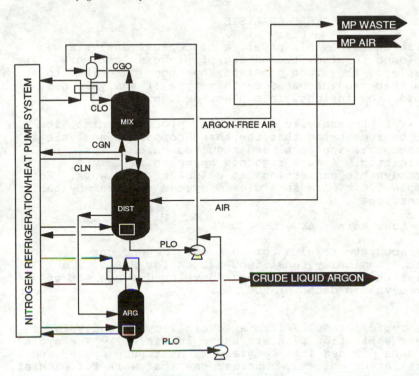

<u>Figure 4</u> Argon Generator Cycle

splits air into oxygen and nitrogen. An argon/oxygen
stream is also withdrawn for purification in a separate
column, from which the argon product stream is
produced. The heat pump and refrigeration requirements
are provided by a nitrogen recycle system, similar to
conventional nitrogen liquefiers.

A mixing column is shown above the distillation
column. Gaseous nitrogen from the distillation column
is passed into the mixing, column, and the mixing column
returns a liquid stream which reduces the reflux which
is required from the nitrogen heat pump circuit.
Similarly liquid oxygen is fed to the top of the mixing
column, and a gaseous oxygen stream is returned. A
mixed waste stream is withdrawn from the centre of the
mixing column.

5 OVERALL CYCLE ANALYSIS

At the start of this paper, a conceptual analysis was
developed to study the nature of process lost work. The
example of the argon generator has been described to
show that studies based on this type of analysis can
provide new insights into plant design.

For the analysis of existing process cycles, it
must be remembered that the three component analysis is
a conceptual approach, and requires care in its
application. A more rigorous approach is to use
thermodynamic expressions to calculate lost work. For
example, lost work in a closed system is given by the
expression:

$$\text{Lost Work} : \Delta X = \Delta H - T_o \Delta S$$

Accurate results for lost work depend on
thermodynamically consistent tables for H & S. To
assist this study S.C. Hwang[3] developed a new set of
thermodynamic data for BOC.

Typical results from an overall cycle analysis of
an air separation plan are shown in Table 2. The cycle
has been divided into compressor, heat pump cycle, and
distillation columns. Whereas the lost work for each of
these components is similar, the efficiency for
distillation is significantly lower than the others.
However we have found that in the argon generator cycle,
much higher distillation efficiencies were achieved.
Therefore we focussed on distillation, to see how high
efficiency could be achieved within the low pressure air
separation cycle.

TABLE 2   Typical Air Separation Unit (1982)

	Input MW	Lost Work MW	Efficiency
Compressor Power	10.0	2.75	72.5%
Power to Heat Pump Circuit	7.25	2.71	62.6%
Net Power to Column System	4.54	2.44	46.1%
Separation Work in Output streams – ex columns	2.10		
Separation work in Product Stream	1.73		

Analysis of Conventional Air Separation Cycles

As has already been highlighted elsewhere in the paper, the distillation columns are clearly the key part of any cryogenic air separation process. The diagram shown on figure 5 illustrates the streams entering and clearing the column system.

The basic system is that air enters the lower column at around 6 bar, and is converted to oxygen and nitrogen at low pressure. The objective is clearly to reduce the work input for separation as much as possible, and there are essentially two possible methods for achieving this in a conventional way;

(1)   Some high pressure nitrogen can be withdrawn from the high pressure column. In this way some of the pressure energy supplied to the process is recovered. The cost of doing this, is, however, that less poor liquid is avaiable and so the reflux ratio in the low pressure column is reduced.

(2)   Some low pressure (generally termed Lachmann) air can be introduced to the low pressure column. In this way the work supplied to the column system is reduced. This also reduces the reflux ratios in the low pressure column, although different sections are affected slightly differently to the above case.

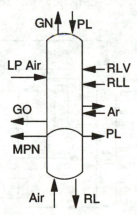

<u>Figure 5</u> Streams Around Double Column

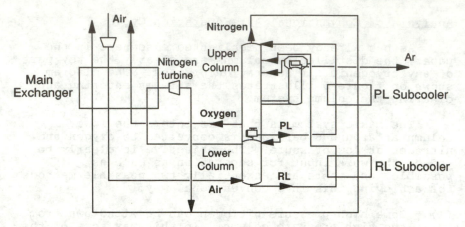

<u>Figure 6</u> Nitrogen Expansion Cycle

        In practice, there are two variants on the double
column system, based on the two cases described above.
Firstly a nitrogen expansion cycle, which is shown in
figure 6. The refrigeration needs of this plant are
provided by removing some nitrogen from the lower
column, warming in the main exchanger, and expanding to
low pressure. The alternative, known as an air
expansion cycle is shown on figure 7. A proportion of
the feed air is expanded in a turbine to low pressure.
This air can then be processed in the low pressure
column, or else bypassed into waste nitrogen stream. A
degree of optimisation on reflux ratios is thus possible
with air expansion plants, which can not be achieved
with nitrogen expansion.

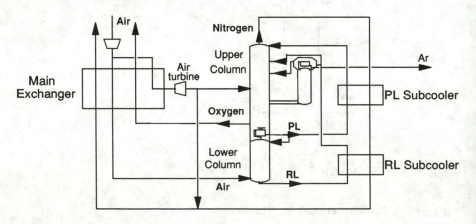

<u>Figure 7</u> Air Expansion Cycle

In addition, there is a degree of flexibility in
that the amount of Lachmann air processed can be varied
according to the product demands at any given time.
However, even with an air expansion plant, only a
limited degree of flexibility is possible, as the reflux
ratios in different sections of the upper column are
still linked to each other, and this limits the degree
of optimisation that is possible.

The above point can be illustrated by considering
the effect of reflux ratio on the efficiency of
different sections of the low pressure column, which is
the part of the process where the bulk of the separation
work is performed. The variable of interest in
performing this exercise is the ratio of liquid to gas
flowates on the column (LN). The thermodynamic
efficiency is plotted against reflux ratio for top and
bottom sections of the low pressure column on figures 8
and 9.

1)   A nitrogen expansion cycle
2)   An air expansion cycle, where all the turbine air
      is processed in the low pressure column.
3)   An air expansion cycle where the turbine air
      bypasses the column system.

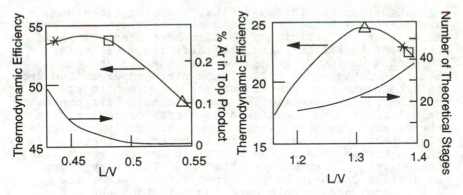

✳ Nitrogen Expansion

☐ Air Expansion, Full Lach.

△ Air Expansion, Zero Lach.

Figure 8 Effect of Reflux   **Figure 9** Effect of Reflux
Ratio at Top of Upper Column   Ratio at Bottom of Upper Column

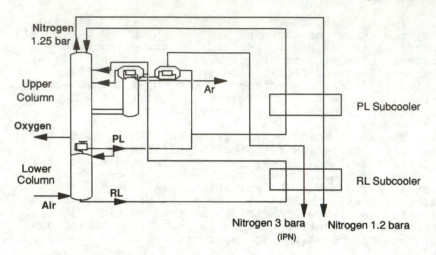

<u>Figure 10</u> Improved Low Pressure Cycle

There are two possible operating extremes for any
distillation column section; minimum reflux and total
reflux. At total reflux, the efficiency of the section
is zero because no separation work is performed, yet
some losses are still incurred.

Equally, at minimum reflux, the pressure drop
associated with the infinite number of trays required
means that the efficiency is zero here also. At some
intermediate reflux ratio, the efficiency will pass
through a maximum. It can be seen that in the top
section of the upper column, the nitrogen expansion and
full Lachmann air expansion cases are close to the
optimum position, while the zero Lachmann, air expansion
case suffers from a cosiderable excess of reflux. In
the nitrogen expansion case, however, conditions are so
tight that argon loss inevitably becomes a factor. In
the bottom section of the column (below the argon draw),
the position is somewhat different. This very tight
separation can only be performed with a rather low
efficiency, and in this case the zero Lachmann air
expansion plant has the highest efficiency. The other
two plants suffer from the pressure drop associated with
the large number of trays required in this section.

In practice, this would probably mean that a lower
purity oxygen would be provided, and so an increased
amount of argon would be lost in the oxygen product.

The above exercise illustrates the point that it is impossible to optimise different sections of the upper column simultaneously with conventional circuitry, and shows a possible weakness which would be improved upon with a suitable modification.

6 AN IMPROVED EFFICIENCY AIR SEPARATION CYCLE

As has been shown above, if a zero Lachmann air expansion plant is considered as a base case, then the top section of the low pressure column is operating with a degree of surplus reflux, while reducing the reflux further down the column is detrimental to the process.

A flowsheet for achieving optimum conditions in both sections simultaneously is shown on figure 10.

A proportion of the poor liquid flow is diverted to the argon condenser and used to provide a proportion of the condensation duty. As a result, less rich liquid is evaporated, and the reflux removed from the top section of the column is returned further down, with the result that conditions in the bottom section remain very close to those in the base case. Thus, a more efficient distillation column has been found than is possible with conventional circuitry.

The way in which the increased efficiency is realised is that the nitrogen in the stream evaporated in the argon condenser (Intermediate pressure nitrogen or IPN) is returned from the column system at a pressure close to 3 bar. The question then arises as to how best the savings available from this arrangement can be realised.

The simplest way for achieving this is shown in figure 11. The IPN stream is simply taken through the main exchanger and the pressure contained in the stream can be credited against nitrogen product compression. With this arrangement the power savings on nitrogen

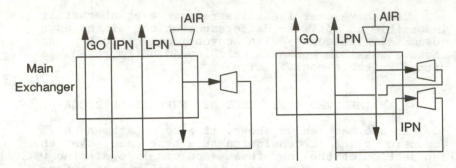

Figure 11 Nitrogen   Figure 12
Produced at Pressure   Twin Turbine Arrangement

compression would amount to around 5% of the air
compression power.

     An alternative arrangement is shown on figure 12.
Two turbines are used, one using air as the working
fluid, and the other intermediate pressure nitrogen.
Since some of the refrigeration requirement is now
provided by the nitrogen stream, less flow is required
through the air turbine, and so the power of the main
air compression is reduced. There is an additional
benefit of this arrangement in that refrigeration can be
provided more evenly along the temperature of the main
exchanger. This is equivalent to saying that the
temperature of the air turbine may be raised, increasing
its production of refrigeration, so that less flow is
needed. Accordingly, with this type of arrangement a 6%
power saving is possible.

            7 AIR SEPARATION AND OTHER PROCESSES

One area which is likely to see development in the
future is the consideration of air separation in a wider
context. Variables such as oxygen purity and pressure
can only be optimised by considering oxygen making and
oxygen using processes as separate parts of a whole.
Another aspect to this is, as has already been
discussed, air separation is a large consumer of power.
Since oxygen using processes are usually, by their
nature, highly exothermic, they are likely to be in
heat/power surplus. Thus there is likely to be some
scope for integrating the process into the heat/power
system of the site.

     One example of where this has been proposed is in

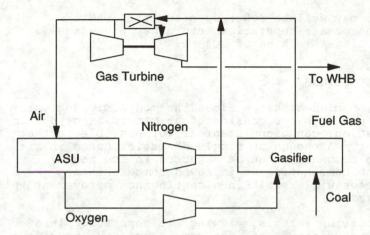

<u>Figure 13</u> Integrated Coal Gasification Power Station

coal gasification/combined cycle electricity
generation[4]. The proposed integration scheme for this
is shown on figure 13. The scheme has a number of
advantages:

1) Nitrogen is used to lower flame temperatures in the
    gas turbine combustion chamber, which gives a good
    method of controlling NO_x formation.

2) The use of nitrogen for this purpose, as well as the
    combustion of a relatively poor fuel gas results in
    the need to bleed air from the machine.

3) The bled gas turbine air could be used a feed for
    the air separation plant. The combination of
    nitrogen returned to the machine, and air bled from
    it results in the right balance between compression
    and expansion on the gas turbine machine.

4) The air separation plant is operated at higher than
    normal pressures. The air feed will be at typically
    10 bar, and the oxygen and nitrogen will be produced
    at a pressure close to 3 bar. The elevated
    pressures mean that the impact of pressure drop
    losses is greatly reduced, and the size of cold box
    is reduced also. Thus, it is a higher efficiency,
    lower capital design than a conventional plant.
    This is only possible, however, because of the use
    for the nitrogen in the gas turbine.

There may well be other opportunities where integration of air separation into other processes is possible using this kind of scheme.

8 CONCLUSIONS

It can be concluded that a close thermodynamic analysis of a process, even one considered as mature as cryogenic air separation, can pinpoint areas for possible further improvement. Although no single dramatic change in air separation technology can be expected in the near future, the combination of improved components and better cycles will result in a continuing improvement in the process.

In addition to this, viewing air separation in a wider process context can, for some processes, provide good opportunities.

REFERENCES

1.   B. Linnhoff et al 'A User Guide on Process Integration for the Efficient Use of Energy', Technical Report, The Institution of Chemical Engineers, 1982.
2.   T. D. Atkinson, Gas Separation and Purification, 1987, 1, 84.
3.   S. C. Hwang, Fluid Phase Equilibria, 1987, 37, 157.
4.   R. Mueller, J. Karg, ASMB/IEEE Power Generation Conference, Miami Beach, Florida, October 1987

LIST OF SYMBOLS USED

H — Enthalpy
P — Pressure
Q — Heat Flow
R — Gas Constant
S — Entropy
T — Absolute Temperature
To — Reterence Temperature
x — Mole Fraction
X — Exergy

Gas Separation by Distillation

K. E. Porter

SEPARATIONS RESEARCH GROUP, DEPARTMENT OF CHEMICAL ENGINEERING AND
APPLIED CHEMISTRY, ASTON UNIVERSITY, BIRMINGHAM B4 7ET, UK

ABSTRACT

Distillation, the most frequently used method of
separating fluid mixtures, is a large mature business,
so that any research and development which changes the
technology may have important commercial implications.
An example is the new market for structured packings
which has developed since the recent research on the
scale-up of packed columns.

A simple explanation is provided for the success of
the empirical methods which are still being used to
design distillation columns. Much is known but little
is understood; thus there is still the potential for
valuable innovation from continuing research. It is
suggested that alternative separation methods such as
adsorption or membranes may sometimes replace cryogenic
distillation but are unlikely to replace pressure
distillation.

1. INTRODUCTION

Distillation is the most frequently used method of separating fluid mixtures on a commercial scale[1]. The dominance of distillation over all the other separation methods is such that the other methods are sometimes reviewed as "Alternative Separation Processes"[2], (alternative, that is, to distillation).

The separation of a fluid mixture by distillation depends on there being a difference in concentration between the mixture when it exists in the form of a vapour and when it exists in the form of a liquid, when these are in equilibrium. The use of the countercurrent contacting of a rising vapour and a falling liquid can usually produce essentially complete separation of the mixture. This is illustrated in Fig. 1, and is described in many undergraduate text books. (e.g. those by Coulson & Richardson or by Treybal).

The main cost of distillation is that of the energy used to boil the liquid at the bottom of a column to provide the vapour which must be condensed at the top. However, distillation columns are particularly suitable for energy integration into the process of which they form a part, or for integrating one column with another[3]. This may much reduce the cost of distillation. In integrated processes, heat loses temperature as it is reused. Heat is said to be degraded from expensive high temperature, high grade heat to cheaper low temperature, low grade heat. Thus in passing through the distillation column heat entering the reboiler is degraded before it leaves the condenser. To quote (approximately)

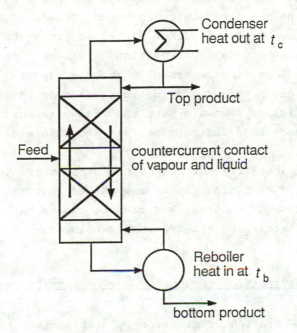

Column temperature difference $\Delta t = t_b - t_c$

Easy separation $\Delta t > 10°C$ Small heat
requirement

Difficult separation $\Delta t \sim 2°C$ Large heat requirement

<u>Figure 1:</u>   A distillation column

Linnhoff "Distillation columns use up temperature
rather than energy".

The design of the distillation column depends on
the choice and sizing of the column internals which
provide the means of achieving vapour-liquid contact
within it. These are either plates or trays (the
terms are synonymous), or packings. Plates support a
moving pool of liquid through which the vapour
bubbles, and packings provide surfaces over which the
liquid trickles downward past the rising vapour.
Examples of trays and packings are shown in Figs. 2a
and 2b.

This paper first discusses the present "state of
the art" of designing trayed and packed columns for
distillation in general, and then considers the
special features of gas separation columns.

2. DISTILLATION TECHNOLOGY - THE STATE OF THE ART

The separation of mixtures by distillation has
been practised for several hundred years, but the
development of modern distillation technology has
coincided with the development of the oil and
petrochemical industry over the last seventy years.
The manufacture and supply of distillation equipment
may then be described as a mature business. Thus it
is often claimed (particularly by specialists in other
separation methods) that all is known that needs to be
known and that therefore no more financial support for
research in distillation is required.

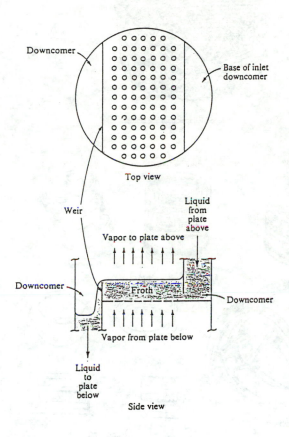

<u>Figure 2a:</u>   Vapour and liquid contacted on a distillation plate (or tray)

Single Mellapak packing element.

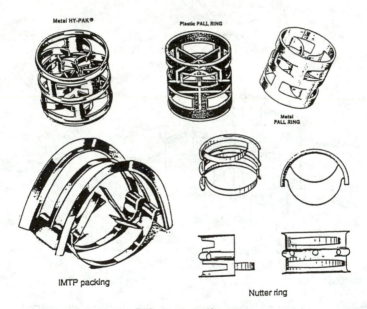

Different random packing types.

Figure 2b: Packings used in distillation columns. A structured packing, Mellapak
is shown above a selection of random packings

However, as with other mature businesses, technical changes are continuous in response to changing economic pressures and market forces. For example, the mature business of brewing now requires separation technology (distillation?) to produce alcohol free beer and wine, and that of manufacturing motor car engines requires new technology to reduce pollution. Small changes in a mature business may have significant commercial implications. The business of distillation has changed significantly in the last five years in response to a requirement to revamp columns in energy saving schemes. This has created a new market for the new structured packings which has been estimated to be about $20 million p.a. in Europe alone. Those changes have come about as a result of research into the design of the liquid distributors which feed the liquid onto the top of the packed bed[7]. This research work may well have had a larger practical and commercial impact than most of the research on the alternative separation methods put together.

It will become apparent from the several examples described below, that we have very little scientific understanding of the phenomena which which determine the performance of trays and packings. This is indeed surprising considering the maturity of the business. Despite this lack of scientific knowledge, it is nevertheless possible to make reliable designs of distillation columns for a wide range of duties, based on previous experience. The explanation for this anomaly was presented in 1979 by Porter and Jenkins[5] and provides a helpful introduction to a discussion of the use of distillation for the separation of gases.

3. EMPIRICISM AND SCIENTIFIC UNDERSTANDING IN THE
DESIGN OF DISTILLATION COLUMNS

To design a distillation column, (given that the
vapour and liquid flow-rates are specified), requires
us to choose the dimensions of the column and of the
column internals such that it will accommodate the
specified flow-rates and achieve the separation. We
must choose the optimum diameter and height of the
column and, if trays are used, share the column cross
sectional area between that part of the tray required
for the vapour flow and that for the liquid flow.
Figs. 3a - 3c and equation 1 are typical of the
relationships used to design columns for specified
throughput (i.e. to choose the column diameters).
They reproduce experience obtained with columns
working at various pressures between say 0.3 bar and
20 bar. In general the design methods assume that the
required dimensions are a function of a) the vapour
and liquid flow rates, and b) the vapour and liquid
densities.

$$\hat{u}_D = 0.07 \, (\rho_L - \rho_V)^{1/2} \tag{1}$$

Figs. 3a, 3b, and 3c, are flooding curves for
packing and trays. Fig. 3c shows the effect of tray
spacing on flooding.

Equation 1 is used to calculate the area of the
downcomer through which the liquid flows between
trays.

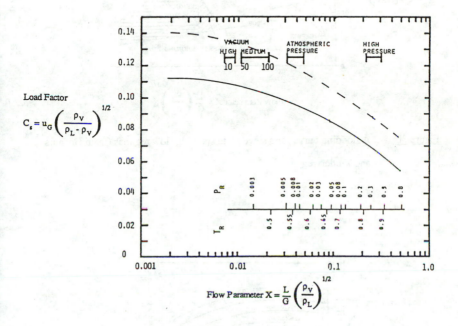

Load Factor

$$C_s = u_G \left(\frac{\rho_V}{\rho_L - \rho_V} \right)^{1/2}$$

Flow Parameter $X = \dfrac{L}{G} \left(\dfrac{\rho_V}{\rho_L} \right)^{1/2}$

<u>Figure 3a:</u>   A maximum capacity of flooding curve for 50mm Pall ring packing.

Full curve 80% flood line

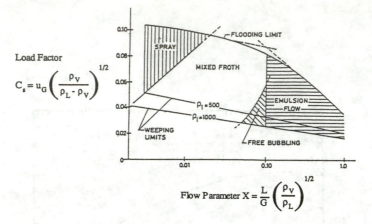

Load Factor

$$C_s = u_G \left(\frac{\rho_V}{\rho_L - \rho_V} \right)^{1/2}$$

Flow Parameter $X = \frac{L}{G} \left(\frac{\rho_V}{\rho_L} \right)^{1/2}$

Figure 3b: A flooding curve for a sieve plate showing flow regimes (from Hofhuis and Zuiderweg).

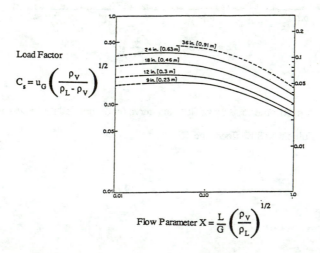

Load Factor

$$C_s = u_G \left(\frac{\rho_V}{\rho_L - \rho_V} \right)^{1/2}$$

Flow Parameter $X = \frac{L}{G} \left(\frac{\rho_V}{\rho_L} \right)^{1/2}$

Figure 3c: The effect of tray spacing on sieve plate flooding.

It is remarkable that only the densities are used in these empirical design methods. We might reasonably assume that some of the other physical properties would have an effect on the capacity of a distillation column. For example, the liquid viscosity might influence the rate of liquid flow down the column, or the interfacial tension might influence the stability of the froth formed on the tray which is expected to separate in the downcomer.It was shown by Porter and Jenkins[5], that for the conditions of practical distillation, (ranging from the vacuum distillation of high molecular weight materials to the distillation of low molecular weight materials under pressure), that all of the physical properties may be correlated against the ratio of the vapour density to that of the liquid density. This is shown in Fig. 4 which is based on the distillation of the substances listed in Table 1. This means that (for the conditions of practical distillation) all of the physical properties correlate one with another so that a given change in interfacial tension is associated with a given change in liquid viscosity or in liquid density or in vapour density. Thus, simple empirical design equations, such as Eq. 1, may be valid expressions of more complex relationships involving the other physical properties. For example, it was shown that a frequently quoted method of predicting tray efficiency, (the AIChE Bubble Tray Design Manual[6]), which consists of 8 equations involving 6 physical properties and 2 flow rates may be represented, <u>for one particular design of tray,</u> by a simple relationship in only the vapour and liquid density, Eq. 2. The above argument also to some

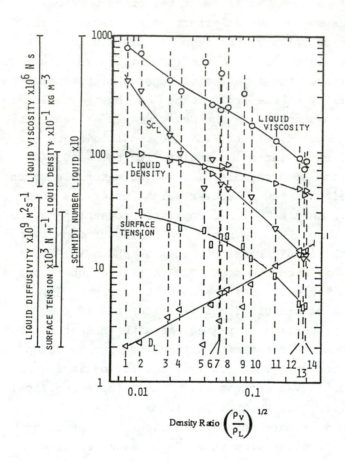

<u>Figure 4:</u>   Physical properties plotted against the density ratio for several systems

at practical distillation pressures.

Table 1   Compounds used to construct Figure 4,
           which shows how the physical properties
           vary with the ratio of vapour and liquid
           densities

Key	Compound	Distillation Pressure (atm)
1	Diphenyl	0.0132
2	Naphthalene	0.0263
3	Ethyl benzene	0.0790
4	p-Xylene	0.132
5	n-Heptane	0.352
6	Cyclohexane	0.352
7	Acetone	1.0
8	Toluene	1.0
9	Cyclohexane	1.69
10	n-Pentane	2.0
11	n-Pentane	5.0
12	isoButane	11.6
13	Ethane	17.8
14	Propane	15.9

extent explains one of the earliest empirical
correlations of tray efficiency, that of Drickamar and
Bradford[10]

$$E_{mv} = 150 + 22 \ln (\rho_V/\rho_L)^{1/2} \qquad\qquad (2)$$

Note, that for an optimum design, at say 80% of
maximum capacity, the design flow rates used with a
particular device also correlate with the density
ratio as shown in Fig. 5 for 50mm Pall rings. This
is because, in distillation, the ratio of liquid flow
to vapour flow is usually close to unity, i.e. 0.8 <
L/G < 1.2. Thus the Flow Parameter is approximated
by the square root of the density ratio. Thus low
vapour density vacuum distillation, represented on
the left hand side of Fig. 5, is associated with a
low liquid rate per unit cross sectional area of
column, and high vapour density pressure distillation
on the right hand side, with a high liquid rate per
unit cross section.

To summarise, in distillation column design it is
a common observation that, in general, at a
particular operating pressure we frequently find a
particular combination of flow rates and physical
properties. This makes the subject an empiricist's
dream and a researcher's nightmare. The empiricist
finds that almost any equation will form a
satisfactory basis for a correlation of previous
experience. The researcher finds it difficult to
separate the effect on column performance of the
different variables, particularly if he bases his
ideas on laboratory experiments at ambient
conditions. An example of this was presented by

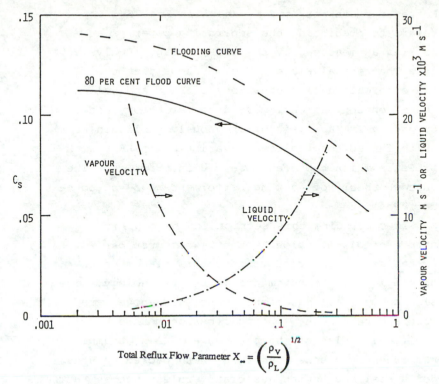

Total Reflux Flow Parameter $X_\infty = \left(\dfrac{\rho_V}{\rho_L}\right)^{1/2}$

<u>Figure 5:</u>   Changes in vapour and liquid superficial velocities with the vapour to liquid density ratio

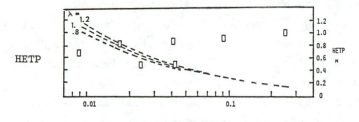

Flow Parameter

<u>Figure 6:</u>   Changes in the H.E.T.P. of 50mm Pall rings with the ratio of vapour density to liquid density. The small rectangles show experimental data. The dotted lines show the predictions of the theory of Bolles and Fair.

Porter and Jenkins[5] and is shown in Fig. 6. It
refers to predicting the separating power of a
packing as measured by the HETP, (Height Equivalent
to a Theoretical Plate). Fig. 6 shows that the
"theoretical" design equations of Fair and Bolles[8],
(based on conventional mass transfer theory and a
comprehensive analysis of previously published work),
quite failed to predict the variation of HETP with
pressure which had been observed in practice. The
unknown theory of packing performance has yet to be
derived.

In concluding this section on the empirical
nature of distillation knowhow, what must be
emphasised is the restrictions this places on the
development of improvements in the technology. The
simple empirical equations, described above, are
complemented in the empirical design method by
empirical "rules of thumb". An excellent example is
the recently discredited rule that "packed columns
don't scale up", with associated rules such as "HETP
= Column diameter". These were based on many
unexplained scale up failures. They effectively
prevented the use of large diameter packed
distillation columns until relatively recently. The
need to conserve energy, (by providing a column with
more theoretical plates so as to use less reflux),
provided the incentive for the research work on
liquid distributors. During the past five years,
this has so changed the technology that packed
columns of large diameter are now designed with
confidence. The advantages to the industry have been
demonstrated by the growth in the market for
packings, described above.

Somewhat provocatively one might describe the present state of the art of the design of distillation columns as more similar to that used in the design of York Minster (~1200 AD) than to that used in the design of the Sydney Opera House. This must be borne in mind when considering the special case of the separation of gases by distillation.

Gas Separation

A gas is here considered to be any substance with a vapour pressure greater than ambient pressure at ambient temperature. In discussing gas separation by distillation, it is necessary to distinguish between high pressure distillation in which the vapour may be condensed by ambient temperature cooling water, and the more expensive cryogenic distillation which is carried out at a reduced temperature. The distinction depends on the critical temperatures of the gases to be separated. This must be above say 50°C (40°C?) to permit condensing with cooling water. The relatively few substances with a critical temperature below 50°C are listed in Table 2. There are far more gaseous substances with a critical temperature above 50°C which may be separated by pressure distillation. They include the hydrocarbons which form both natural gas and synthetic natural gas, both of which will already be available under pressure before distillation, and halogenated hydrocarbons (e.g. refrigerants). The technologies of pressure distillation and cryogenic distillation are sufficiently different to merit separate treatment.

Table 2   Gases with a Critical Temperature
           less than 50°C (323°K)

	Name	Critical Temp °K
	Helium	5.2
	Neon	44.4
	Argon	150.8
	Krypton	209.4
AS	Hydrogen	33.2
AS	Nitrogen	126.2
	Nitric Oxide	180.0
	Nitrous Oxide	309.6
AS	Oxygen	154.9
AS	Carbon Monoxide	132.9
AS	Carbon Dioxide	304.2
AS	Methane	190.6
	Ethane	305.4
	Ethylene	282.4
	Acetylene	308.3
	Carbontetrafluoride	227.6
	Chlorotrifluoromethane	302.0

AS =   Gas sometimes separated by Adsorption or
       Membranes.(Alternative Separation Method)

Table 3   Some Empirical "Foam Factors"
           (used to derate distillation column
            design, i.e. increase column diameter)

Name	Foam Factors (Approximate)
Methane	0.7
Ethane	0.8
Propane	0.9

Pressure Distillation

The column is maintained at a pressure at which
the top product may be condensed by cooling water.
To allow a sufficient temperature driving force for
heat transfer, this may be taken to be 50°C. Even
allowing for a temperature rise over the distillation
column of 10 or 20°C to allow for the change in
composition, the temperature required for the heat to
be supplied to the reboiler is relatively low.
Pressure distillation requires low grade low cost
heat. Distillation columns are well suited for
energy integration, and if the pressure distillation
column forms part of a larger process, it may well be
possible to use waste heat for the column reboil, in
which case the energy costs will be small.

Compared to distillation at lower pressures, the
column capital cost per mass of product is less than
might be expected. This is because the throughput
increases with pressure, and may be as high as 40,000
kg/hr of vapour per m^2 of column cross sectional area
at 10 bar, compared with say 15,000 kg/hr m^2 at 1
bar. These rates would be for a tray spacing of
600mm. The large froth heights associated with the
high liquid rate, and the low vapour velocity
associated with the high vapour density, result in
high tray efficiencies approaching 100%; thus
pressure distillation is a relatively cheap operation
and it seems unlikely that it will be replaced by the
alternative separation methods.

The design of pressure distillation columns forms
part of the empirical knowledge which has developed in
the oil refineries and petrochemical plant.

Traditionally trays are used at tray spacings of 450mm
to 600mm so that they may be installed into the
columns on site. Often these are sieve trays with
12.5mm diameter holes. The high vapour rates used at
these high tray spacings are associated with high
liquid rates, and a large proportion of the column
cross sectional area is occupied by the liquid
downcomers. Columns tend to be multipass with a
relatively short length of liquid flow path.

It seems that the empirical design procedures
developed from experience at lower pressures have
sometimes failed at higher pressures. That is, high
pressure distillation columns have sometimes been
underdesigned and have not accepted the specified
throughput. It is not uncommon to allow for this by
using special empirical "foam factors" to reduce the
design velocities calculated by the traditional
procedures. Examples of foam factors are given in
Table 3. These "foam factors" are no more than
empirical derating factors. They do not necessarily
imply that with high pressure systems the column fills
with foam. Zafir Ali[9] has shown that the same system
which produces a foam in a 50mm diameter glass and
sintered disk apparatus may produce a massive
entrainment of fine drops, but no foam, in a pilot
scale tray test rig.

It has been suggested that this unexpected loss in
throughput may be due to a change in the mode of gas-
liquid contacting at high pressure, to the régime of
"emulsified flow"[10]. In this régime, froth leaving
the tray flows over the outlet weir into the downcomer
as if it were a single phase liquid. At lower
pressure, in the "mixed" or "spray" régimes, liquid

transfer into the downcomer is by splashing or by drops in free flight. The different flow phenomena associated with the emulsified flow régime may cause a) unexpected choking of the downcomer inlet, or b) massive vapour entrainment through the downcomer to the tray below[10]. It is not yet clear what determines downcomer flooding at high pressures or what is the scientific explanation for the empirical foam factors.

In recent years, packings have sometimes been used in pressure distillation. At first these were randomly packed (for example the new metal saddle shapes such as Intalox Metal Tower Packing, first described by Strigle and Porter[11]). More recently structured packing has been used (for example Sulzer Mellapac[12] or Norton Structured Intalox). The HETP of packings in high pressure distillation has been found to be higher than that observed at lower pressures[13]. This unexpected loss in separating power cannot be explained by existing theories. It may be due to vapour back mixing, or to some change in the liquid flow pattern, associated with the higher rate of flow, or low surface tension of the liquid.

Cryogenic Distillation

The technology of cryogenic distillation has developed differently from that of high pressure distillation. The largest business is the separation of air where plants have been built to supply as much as 2,000 tonnes per day of oxygen with distillation columns up to 6 m in diameter operating in the temperature range of 77 to 90°K. The usual practice is to minimise the heat transfer from the surroundings

into the plant by enclosing it inside an insulated
"cold box". Shorter, fatter columns are preferred to
longer, thinner ones, so as to reduce the surface area
of the column and the cost of the cold box. This has
led to the practice of close tray spacings of about
150mm, which is much lower than that used in other
distillation columns. Trays are installed into the
column in a workshop before the column is sent to the
site. Sieve trays with small holes of 1 - 2mm
diameter are often used, and low pressure drop designs
are preferred on economic grounds.

The close tray spacings and low pressure drops
result in a lower vapour flow per unit cross
sectional area of column. Thus most of the column
cross section is used for vapour flow and the result
is large bubbling areas with long liquid flow paths.

The cost of cryogenic distillation includes the
liquefaction of some of the gas by the expansion of
cold gas through a nozzle or turbine and the
provision of large heat exchangers to cool the liquid
feed gas by warming the products. Additional costs
are those for removing any parts of the mixture which
will solidify in the plant at cryogenic temperatures,
such as water and carbon dioxide, or cause a risk of
explosion such as acetylene.

Thus whereas pressure distillation is relatively
cheap, cryogenic distillation is relatively
expensive. Applications of the newer alternative
methods of separation have so far been for mixtures
which would require their separation by distillation
to be cryogenic. Table 2 lists some commercial
applications of gas separation by membranes or by
pressure swing adsorption. (The gas mixtures in this

table are those with a critical temperature less than 50°C)

Some of the problems in the design of distillation plates for cryogenic distillation are a result of the large bubbling areas and long liquid flow path lengths mentioned above. It is believed that some of the companies supplying air separation plant continue research into the liquid flow patterns. The liquid flow pattern can have a significant effect on the separating power of the column. This has been demonstrated in recent work at Aston University. The flow pattern effects only show up in large diameter columns, and the Aston work is unique in that we use a tray test rig 2.4m in diameter.

We simulate distillation by cooling hot water with air and measure the water temperature at over 100 points on the tray. Water temperature is analogous to the liquid concentration in a distillation tray. A coloured dye is injected on to the tray at the water inlet and the nature of the liquid flow pattern is observed by following the progress of the dye across the tray. This is recorded on video.

Two quite different flow patterns may be observed, determined by the rate of flow of the liquid. At low flow rates, after leaving the inlet downcomer, the liquid spreads out across the width of the tray and then converges towards the outer weir producing a large slow moving region in the middle of the tray near the outlet weir. At high liquid flow rates, the liquid crosses the tray from inlet to outlet without spreading out, forming circulating

eddies at the sides of the tray. This form of liquid
channelling has been responsible for expensive scale
up failures in reduced pressure distillation[14] which
were predicted by the theoretical models of Porter
and Lockett et al[15], (see also 5).

The different flow patterns produce quite
different temperature profiles on the tray, as
illustrated in Figs. 7a and 7b which show lines of
constant temperature. Fig. 7a is a high liquid rate
experiment and Fig. 7b is a low liquid rate
experiment. The sieve plate in these experiments is
similar to those used in cryogenic practice, (i.e. 1mm
diameter holes). It is expected that in distillation,
the liquid concentration profiles will be similar to
the temperature profiles, so that the changes in the
liquid flow pattern will have a significant effect on
column performance.

A scientific understanding of these phenomena
requires a theory of open channel two phase flow which
will account for the influence of the momentum
exchange between the vapour emerging through the sieve
plate holes and the liquid. The development of this
theory is part of our research programme. The
empirical design methods take no account of how these
liquid flow patterns influence tray efficiency or of
how they might be controlled by tray design based on a
scientific understanding of two phase flow.

In this mature business, although much is already
known, little is understood, and only by more research
can we hope to be able to produce new, improved,
scientific designs, free from an empiricism based on
avoiding the failures of the past.

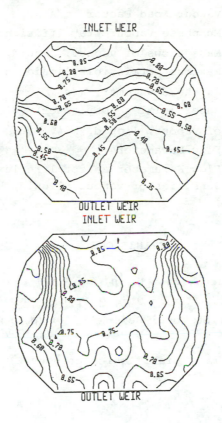

<u>Figure 7:</u>   Water cooling on a 2.4m diameter sieve plate. Lines of constant

temperature (isotherms) are shown. Figure 7a is for a high liquid flow rate and

Figure 7b for a low liquid flow rate.

NOMENCLATURE

C_S	=	Vapour Load Factor, m/s.
E_{mv}	=	Murphree Vapour Tray Efficiency
G, L	=	Mass flows of vapour, liquid, kg/hr.
$\hat{u}_D$	=	Maximum permissible liquid velocity in the downcomer, m/s.
U_G, U_L	=	superficial (empty tower) velocities of vapour, liquid, m/s.
X	=	Flow Parameter
X_∞	=	Flow Parameter at total reflux
ρ_V, ρ_L	=	Densities of vapour, liquid, kg/m^3.

DEFINITIONS

Vapour Load Factor $C_S = U_G \left(\rho_V / (\rho_L - \rho_V) \right)^{\frac{1}{2}}$, m/s

Flow Parameter, $\quad X = {}^L/_G \, (\rho_V/\rho_L)^{\frac{1}{2}}$

Flow Parameter at Total Reflux $X_\infty = (\rho_V/\rho_L)^{\frac{1}{2}}$

Note that for many distillation columns $X \simeq X_\infty$

REFERENCES

1. C.J. King, <u>Separation Processes</u>, <u>2nd Ed</u>. McGraw-Hill, New York (1980).

2. K.E. Porter and J.D. Jenkins. Alternative Separation Processes, <u>The Chemical Engineer</u> (<u>Distillation Supplement</u>) (Sept. 1987).

3. K.E. Porter, S.O. Momoh and J.D. Jenkins. Some simplified approaches to the design of the optimum heat-integrated distillation sequence. "Distillation and Absorption '87" <u>I. Chem E</u>. <u>Symposium Ser. No. 104</u>, A449 (1987).

4. F.J. Zuiderweg, P.J. Hoek and L. Lahm Jr. The effect of liquid distribution and redistribution on the separating efficiency of packed columns. "Distillation and Absorption 1987" <u>I. Chem. E</u>. <u>Symposium Ser. No. 104</u>, A217 (1987).

5. K.E. Porter and J.D. Jenkins. Interrelationship between industrial practice and academic research in distillation and absorption. "Distillation '79" <u>I.Chem. E. Symposium Ser</u>. <u>No. 56</u>, 5.1/1 (1979).

6. "<u>Bubble-Tray Design Manual</u>, Am. I. Chem. E., New York (1958).

7. H.G. Drickamar, and J.R. Bradford, <u>Trans. Am. I</u>. <u>Chem. E</u>. <u>39</u>, 319 (1943).

8. W.L. Bolles and J.R. Fair. Performance and design of packed columns. "Distillation '79", <u>I. Chem. E. Symposium Ser. No. 56</u>, 3.3/35 (1979).

9. Z. Ali <u>PhD Thesis, Aston University</u> (1989).

10. F.J. Zuiderweg, P.A.J. Hofhuis and J. Kuzniar. <u>Trans. I. Chem. E.</u>, <u>62</u>, 39 (1984).

11. R.F. Strigle and K.E. Porter. Metal Intalox, a new distillation packing "Distillation '79". <u>I. Chem. E. Symposium Ser. No. 56,</u> 3.3/19 (1979).

12.   L. Spiegel and W. Meier. Correlations of the
      performance characteristics of the various
      Mellapak types. "Distillation and Absorption
      1987". I. Chem. E.Symposium Ser. No. 104, A203
      (1989).

13.   K. Robinson. Lecture on Aston Short Course
      "Advances in distillation"
      (1988).

14.   V.C. Smith and W.V. Delnicki, Chem. Eng. Prog.
      71 (8) 68 (1975).

15.   M.J. Lockett, K.E. Porter and K. Bassoon.
      Trans. I. Chem. E. 53 125   (1975).

Membrane Structure and Gas Separations

V. T. Stannett*

RESEARCH TRIANGLE INSTITUTE, PO BOX 12194, RESEARCH TRIANGLE PARK, NC 27709, USA

R. T. Chern

DEPARTMENT OF CHEMICAL ENGINEERING, NORTH CAROLINA STATE UNIVERSITY, RALEIGH, NC 27695, USA

1 INTRODUCTION

The idea of using membranes for the separation of gases was first suggested by Thomas Graham in 1866.[1] He demonstrated the method practically using natural rubber membranes for the oxygen enrichment of air. These observations were preceded by noting that as the thickness of the membrane was decreased the flux increased but the selectivity between gases remained similar. It appears to have taken nearly one hundred years for any large scale pilot plant process to be developed. This process developed by Stern et al.[2] used fluorocarbon membranes to separate helium from natural gas.

The observation by Graham that the flux could be increased greatly without loss of selectivity by reducing the membrane thickness has led to many developments in their use for separation processes. In particular, the use of asymmetric rather than dense membranes became highly developed during research into desalination processes. Two methods were developed: a sophisticated casting method whereby cast films were briefly air dried on the top surface and then immersed in a non-solvent to develop a porous layer. A second technique used was to dip-coat or generate by interfacial polymerization a pore free selective polymer film onto a porous substrate. It is clearly difficult to guarantee the formation of a defect free skin on an asymmetric membrane and

---

* Professor Emeritus of Chemical Engineering
   North Carolina State University

82

a post treatment with e.g. a highly permeable silicone
rubber, was successfully developed to overcome this
problem. A brief but elegant discussion of the asymmetric membrane types together with the necessary references has been presented by Koros et al.[3] It should be
noted that hollow fiber systems with similar properties
have also been developed.

Research is also underway to produce extremely thin
intact coatings of rejecting polymers using plasma
treatments and Langmuir-Blodgett films. The need for
preparing large surface area defect free membranes by
these techniques present a formidable problem but the
rewards could be great if successful. In addition, the
rejecting layers could in principle have unique separation factors.

2 SOME FUNDAMENTAL CONSIDERATIONS

The solution-diffusion mechanism for the transport
of gases through polymer films (membranes) was clearly
described by Graham in 1866. The history since that
time has been summarized by Stannett[4] and need not be
expanded further.

Briefly, the gas dissolves at the membrane surface,
at the upstream side diffuses through by a series of
activated steps driven by the concentration or strictly
the chemical potential, gradient and evaporates at the
low pressure, downstream, side.

The transport equations have been well developed and
described both for rubbery and glassy polymers both for
single and for mixed gases. It is quite clear that the
solubility and diffusivity (mobility) factors are
involved plus the driving force, i.e. the pressure
gradient, itself.

The permeability coefficient P is the normalized
flux i.e. the flux divided by the pressure gradient. If
the downstream pressure is kept close to zero the permeability can be expressed as:

$$P = \frac{flux}{\Delta p/l} = \frac{\int_{o}^{c} D(c)\,dc}{c} \cdot \frac{c}{p} = D.S.$$

where l is the film thickness, D is a concentration averaged diffusivity and S is the apparent solubility coefficient.

When two different gases A and B are involved and the driving force, usually the pressure differential, is held constant, one can define a separation factor

$$\alpha_{AB} = P_A/P_B = D_A/D_B \cdot S_A/S_B.$$

If there is no interference with the process by one gas such as plasticization as observed by Paul[5] and Koros[6] for carbon dioxide in polycarbonate under high pressure and for cellulose esters by Donohue et al.,[7] the ratios of the permeabilities of the pure components, often called ideal separation factors, can be used.

If one considers the solubility and diffusivity values for the many gas-polymer systems which have been reported, it is clear that whereas diffusivities vary over many orders of magnitude in different polymers, the ratios of the solubilities are mainly governed by the nature of the gases themselves. Considering however the wide variety of gases involved, it may not always be the case. With the present state-of-the-art however rather small but possibly significant changes can be achieved in the separation factors by modification in the solubilities themselves. The present paper dealing as it does with the polymer structure of the membrane will however consider both factors.

Since it seems clear that the diffusivities are the chief criteria for designing gas separation membranes, it is of considerable interest to plot the diffusitivities versus the penetrant size for a number of penetrants in two polymers; one glassy, rigid PVC and one natural rubber. The Van der Waals volumes were used as they were available for all the penetrants considered. These plots have been published[8] and are reproduced in Figure 1. Much greater selectivity is shown by the glassy polymer and this may be attributed to the smaller segmental motions involved with glassy polymers leading to the greater discrimination between the size of the penetrant molecule. Essentially all practical gas separation membrane structures are glassy with some possible exceptions where extremely high fluxes such as with the siloxane rubbers are exploited.

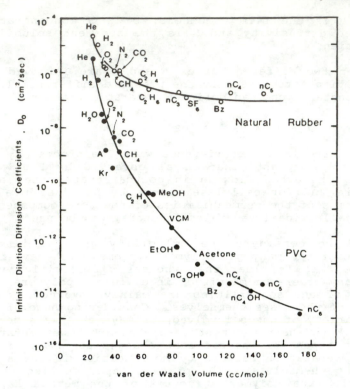

Figure 1.   Diffusion coefficients for a variety of penetrants in
             natural rubber at 25°C and rigid poly(vinyl chloride)
             at 30°C.   The Van der Waals volumes are taken from
             The Handbook of Chemisry and Physics, 35th ed., 1953-
             54, page 21-24 to 21-26, CRC, Cleveland, Ohio.

     The range of gas permeabilities available in polymer
films is illustrated in Table I.[9]   The newly introduced
polymer PTMSP [poly(1-trimethyl(silyl)-1-propyne)] is
about an order of magnitude greater in permeability than
polydimethylsiloxane but the separation factor is within
the range given.   The table therefore shows a spread of
seven orders of magnitude in atmospheric gas permeabilities.   The ratios (ideal selectivities) between oxygen
and carbon dioxide for example only extend about 3-fold.
A similar tendency is shown for oxygen and nitrogen.
This illustrates one of the difficulties in designing
membranes with excellent selectivities particularly for
the atmospheric gases.   Even so, differences in factors
of 3 or so can be of considerable commercial signifi-

cance. The comparative constancy of the ratios was first reported and discussed by Stannett and Szwarc.[10] The permeabilities were attributed to three factors, the nature of the polymer, the nature of the gas and any specific interactions between the two. The latter is often close to unity for many gases and represents comparatively speaking only a correction term. In extreme cases such as water vapor, the interaction term is obviously more important as shown in Table I. Even methane shows some such effects for example in hydrocarbon polymers, more important examples are carbon dioxide in polycarbonates[5,6] and with cellulose esters.[7] There is another highly important breakdown to the rule and that is when extremely low diameter gas molecules are involved such as with helium and hydrogen. The ratios of the permeabilities of a number of polymers, both glassy and rubbery to helium and oxygen are presented in Table II.[8] Similar large selectivity variations can be shown with hydrogen and even with neon. These observations are clearly of great importance in the design of gas separation membranes. They can be attributed to an essentially unknown factor with polymers and that is the actual distribution of the sizes of the fluctuating free volume elements in the membrane.

Pertinent Structure-Property Relationships for D and S in Polymers

Diffusivities. The diffusivity of a given gas in a polymer depends on the number of "holes" in the polymer and the rapidity of interchange or movement among them.

The former depends to a large extent on the free volume or in less sophisticated terms in the packing density of the polymer. This in turn depends on the bulkiness of the segments and intersegmental attraction. The latter depends more on torsional mobility of the chain segments which depend on the rotational energy barrier and, again, on the bulkiness of the chain segments.

Clearly the diffusivities depend also on the size and shape of the gas molecules themselves. With the high pressures used in industrial gas separations, the attraction between the gases and the polymers leading for example to plasticization is also a factor to be considered.

Table I. The Permeability of Various Polymers to Oxygen, Water, and Carbon Dioxide at 30°C.[9] Units: $cm^3(STP)$-$cm/(sec$-cm^2-cm $Hg)$ x 10^{-10}.

Material	$P(O_2)$	$P(CO_2)$	Ratio (CO_2/O_2)	$P(H_2O)$
Polyacrylonitrile	0.0003	0.0018	6.0	300
Polymethacrylonitrile	0.0012	0.0032	2.7	410
Lopac (Monsanto Co.)[a]	0.0035	0.0108	3.1	340
Polyvinylidene chloride	0.0053	0.029	5.5	1.0
Borox (Sohio Co.)[a]	0.0054	0.018	3.3	660
Polyethylene terephthalate	0.035	0.17	4.9	175
Nylon 6	0.038	0.16	4.2	275
Polyvinylchloride (unplasticized)	0.045	0.16	3.6	275
Polyethylene (dens. 0.964)	0.40	1.80	4.5	12
Cellulose acetate (unplasticized)	0.80	2.40	3.0	6900
Butyl rubber	1.30	5.18	4.0	120
Polycarbonate	1.40	8.0	5.7	1400
Polypropylene (dens. 0.907)	2.20	9.2	4.2	65
Polystyrene	2.63	10.5	3.8	1200
Polyethylene (dens. 0.922)	6.90	28.0	4.0	90
Neoprene	4.0	25.8	6.5	910
Teflon®	4.9	12.7	2.6	33
Poly(2,6-dimethylphenylene oxide)	15.8	75.7	4.8	4060
Natural rubber	23.3	153	6.6	2600
Poly(4-methyl pentene-1)	32.3	92.6	2.9	-
Polydimethylsiloxane	605	3240	5.3	40000

[a] High acrylonitrile copolymers.

Solubilities. The solubility of a gas in a polymer depends to a large extent on the force constants of the gases themselves such as the Lennard-Jones force constant, the critical temperature and boiling point or more directly their condensibility. Again the packing density and the Hildebrand solubility parameters must also be involved.

This paper will address itself to the modification of gas separation membranes to develop higher fluxes and improved separation factors between the gases to be separated. Two families of polymers the polyimides and the polycarbonates have been studied intensively in this respect by Koros, Paul and their colleagues. This paper extends such studies but in less detail to the polyphenylene oxides and the aromatic polyesters.

Table II. Permeability of Polymers to Oxygen and Helium at 25°C.
Units: $P = cm^3(STP)\text{-}cm/sec\text{-}cm^2\text{-}cm\ Hg) \times 10^{-10}$.[9]

Material	$P(O_2)$	$P(He)$	Ratio (He/O_2)
Polyacrylonitrile	0.0003	0.53	1770
Polyvinylidene chloride	0.0053	0.31	58.5
Polyethylene terephthalate	0.035	1.32	37.7
Nylon-6	0.038	0.53	13.2
Polyvinyl chloride	0.045	2.05	45.6
Polyethylene 0.964	0.40	1.14	2.15
Polyvinyl acetate	0.50	12.6	25.2
Cellulose acetate	0.80	16.0	20.0
Butadiene-acrylonitrile (61-39)	0.96	6.81	7.1
Polyethyl methacrylate	1.28	18.0	14.0
Polypropylene	2.20	38.0	17.3
Polystyrene	2.63	18.7	7.11
Polyethylene (0.914)	2.90	4.9	1.68
Hydropol (0.894)	11.3	15.7	1.38
Poly(2,6-dimethylphenylene oxide)	15.8	78.1	4.9
Poly(4-methyl pentene-1)	32.3	101	3.13
Polydimethyl siloxane	605	233	0.39

It is clear that successful membranes developed in this type of research can also be, in principle, to the asymmetric processes described earlier and leading to additional improvements. Before proceeding further, attention should be drawn to the really thoughtful and thorough discussion on membrane materials by Koros *et al.*,[3,11] to which attention has already been drawn.

3 EFFECT OF STRUCTURAL MODIFICATION OF POLYPHENYLENE OXIDE AND SOME AROMATIC POLYESTERS ON GAS PERMEABILITY AND PERMSELECTIVITY

In the ensuring discussion, we will focus on how repeat-unit structure affects packing and local chain motions of the polymer, and the consequential changes in gas-transport properties of the polymer. Our objective is to demonstrate how effectively structural modification of the polymer repeat-unit can alter the solubility and diffusivity of carbon dioxide and methane, and therefore the permeability and permselectivity of these polymers to the two gases.

   PPO and Aryl-brominated PPO's.   The first group of
polymers to be covered include an unmodified poly(2,6-
dimethylphenylene oxide), PPO, and three aryl-brominated
PPOs.   Results from these studies had previously been
published.[12,13] Pertinent material data for those polymers are reproduced in Table III.   The Van der Waals
volume used in the calculation of packing density $1/v_f$
in Table III was estimated by the group-contribution
method developed by Bondi.[14] The composition of the
brominated polymers were determined from ^{1}H-NMR spectra
of the polymers.

   The data in Table III clearly show that aryl-bromination is an effective way of reducing the packing density of PPO (as will be shown below, similar statements
can be made for aromatic polyesters).   Moreover, substitution by bromine at the 3,5 positions of the 2,6-
dimethylphenylene oxide entity may be more effective
than mono-substitution in reducing the packing density.
This tentative conclusion is based on the following
argument.   The smaller packing density of PPOBr(0.91)
relative to PPOBr(0.36) clearly establishes the effects
of mono-bromination on packing density since both polymers have roughly the same fraction of 3,5-dibrominated
repeat-unit (see Table III).   On the other hand, the
fractions of un-brominated repeat-unit are roughly the
same in PPOBr(0.91) and PPOBr(1.06), approximately 0.17.
Evidently, the effects of mono- and di-bromination on
packing density must be different, otherwise the packing
densities of PPOBr(0.96) and PPOBr (1.06) should be
similar.   The much smaller packing density of PPOBr
(1.06) than that of PPOBr(0.91), therefore, should be
attributed to the much higher per percentage of 3,5-
dibromination in PPOBr(1.06), unless there is a
synergistic effect resulting from some peculiar combination of mono- and di-bromination.   Since the substitution reaction is presumably random along the polymer
molecule, any synergistic effect should be considered
unlikely.

   The effects of aryl-bromination on the gas-transport
properties of PPOs are summarized in Table IV.   Clearly,
there is a monotonic increase in the apparent solubility
coefficients of both gases as the extent of aryl-bromination is increased.   Since the calculated solubility
parameters of these polymers are essentially the same,
the increase in gas solubility should be attributed to
the reduction in packing density brought about by aryl-
bromination.   A recent study on novel polyimides showed

an even bigger effect of reduced packing on gas solubility.[15]

Table III.   Pertinent Material Properties of the Modified PPO
             Polymers.

	Polymer[*]			
	PPO	PPOBr (0.36)	PPOBr (0.91)	PPOBr (1.06)
% Repeat-Unit Modified				
Mono-brominated	0	20.4	73.9	60.0
Di-brominated	0	7.8	8.5	24.0
Unmodified	100	71.8	17.6	16.0
Density (g/cm^3)	1.061	1.203	1.380	1.390
Packing Density[**] (1/V_f)	2.55	2.47	2.38	2.26
Glass Transition (°C)	210	233	262	269

[*]   The number in the parenthesis represents the average bromine-
      atom/repeat-unit molar ratio in the polymer.

[**] Defined as $\hat{v}/(\hat{v}-\hat{v}_{vw})$ where $\hat{v}$ is the measured specific volume
      at 23°C, and $\hat{v}-\hat{v}_{vw}$ is the calculated specific Van der Waals
      volume of the polymer sample.

     The diffusivities of both carbon dioxide and methane
are highly correlated with the packing density of the
polymer (also see Table III). When the packing density
of the polymer was reduced from 2.55 to 2.47, the diffusivities of both carbon dioxide and methane both
decreased.   But, as the packing density was reduced
further, the diffusivities of both gases showed significant increases.   This behavior can be rationalized by
considering the counteracting effects of packing and
torsional mobility of the repeat-unit of PPO on the diffusivity of penetrant in the polymer.

Table IV.   Representative Values of Permeability P, Permselec-
             tivity P_{CO_2}/P_{CH_4}, Apparent Solubility Coefficient (S),
             and Concentration-average Diffusivity (D), for the
             PPOs at 20 Atm and 35°C.

	Polymer[*]			
	PPO	PPOBr (0.36)	PPOBr (0.91)	PPOBr (1.06)
P				
CO_2	50	51	68	108
CH_4	2.9	2.7	3.4	6.3
[S]				
CO_2	2.22	2.41	2.57	2.76
CH_4	1.08	1.20	1.33	1.45
[D]				
CO_2	17.3	16.1	20.1	29.9
CH_4	2.1	1.8	2.0	3.3
P_{CO_2}/P_{CH_4}	17	19	20	17

All the transport data reported in this paper have the
following units:

$$P = 10^{-10} \text{ cm}^3 \text{ (stp)-cm/cm}^2\text{-sec cmHg}$$
$$[S] = \text{cm}^3 \text{(stp)/cm}^3 \text{ polymer-atm}$$
$$[D] = 10^{-8} \text{ cm}^2/\text{sec}$$

Low temperature mechanical relaxation data have been
widely used for probing local chain motions in aromatic
polymers. Shifting of the gamma-transition peak of both
chlorinated and brominated bisphenol A polycarbonates
was interpreted to mean that aryl-halogenation lessens
local chain motions although the exact nature of the
motions is still uncertain.[16,17] We would like to believe
that the gamma relaxation is associated with pi-flips or
torsional mobility of the aromatic rings. If the preceding interpretation is correct and generally applica-
ble, it appears reasonable to assume that aryl-bromination decreases torsional mobility of the repeat-unit of
PPO. Presumably, a reduction in local chain mobility
such as proposed above will decrease the diffusivity of
a penetrant in the polymer. On the other hand, a reduction in packing is expected to increase the penetrant
diffusivity. The diffusivity data in Table IV can
therefore be interpreted in the following manner: When
the degree of bromination is low, the counteracting

effects of diminished torsional chain mobility and reduced packing on gas-diffusivity are comparable to each other; however, when the packing density is reduced further, the effect of diminished torsional mobility on diffusivity is overwhelmed by the markedly more important effect of reduced packing.

   Aromatic Polyesters. The second group of polymers include polyesters formed by reacting a 50/50 mixture of tere- and iso-phthalates with bisphenols of different structures. The repeat unit of these polyesters are summarized in Figure 2. Pertinent material data of the four polymers are shown in Table V.

Table V. Material Data for the Polyesters.

Polymer	Density at 23°C (g/cm³)	$1/V_f$	δ^{**} $(cal/cm^3)^{1/2}$	Tg (°C)
PAr	1.204	2.77	10.3	191
PPha	1.298	2.78	11.0	279
HFPAr	1.418	2.67	9.4	202
TClPAr	1.368	2.70	10.1	247

* All the densities reported in this paper were determined
    with a density gradient column

** All the solubility parameters reported in this paper were
    calculated from the group contribution method developed by
    Hoy.[18]

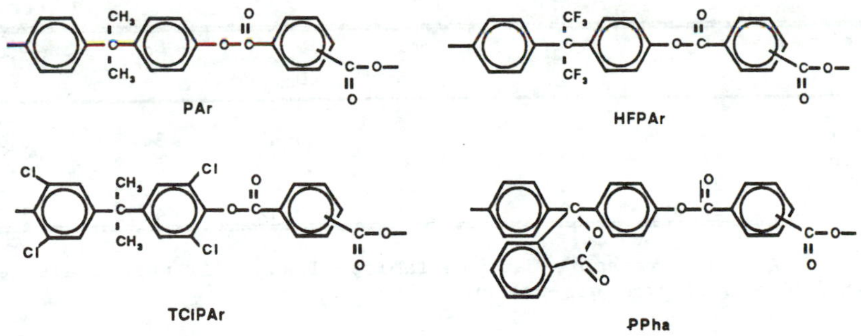

Figure 2. Structures of Polyesters Studied.

With PAr taken as the reference structure, it is clear that substituting 3,3',3",3"' hydrogen atoms of bisphenol A with chlorine brings about a reduction in the packing density. This behavior is similar to the reduction in packing density of PPO due to aryl-bromination discussed in the preceding section. Substituting the isopropylidene unit in the bisphenol A with phenolphthalein does not reduce the packing of the polymer although substitution with hexafluoroisopropylidene results in a remarkable reduction in the packing density. One possible explanation for the ineffectiveness of phenolphthalein in reducing the packing density is an increase in intermolecular attraction because of the presence of many additional polar ester groups in the polymer.

Similar to the PPO system presented earlier, the solubility and diffusivity of gases in the above polyesters are notably affected by the changes in packing density and local chain motions. A summary of the gas-transport data are presented in Table VI.

The larger carbon dioxide permeability of PPha relative to PAr is apparently due to the larger solubility of carbon dioxide in PPha, rather than differences in the carbon dioxide diffusivity. Since PAr and PPha have essentially the same packing density, the similarity in carbon dioxide diffusivity is not unexpected.

Table VI.   Permeability, Diffusivity, and Solubility Data for the Polyesters.

Polymer	P_{CO_2}	P_{CO_2}/P_{CH_4}	D_{CO_2}	D_{CO_2}/D_{CH_4}	S_{CO_2}	S_{CO_2}/S_{CH_4}
PAr	9.0	19	2.78	6.42	3.23	3.10
PPha	14.5	27	2.74	7.33	5.29	3.47
HFPAr	25.0	24	5.44	7.03	4.59	3.42
TClPAr	11.9	29	2.42	9.42	4.92	3.09

All the transport data in Tables VI and VIII were evaluated at 35°C and 10 atm. gas-pressure

Compared with PAr, HFPAr has a significantly smaller packing density. Following the same argument presented earlier, if one may extrapolate conclusions drawn from low temperature mechanical relaxation data of bisphenol

A polycarbonate and hexafluorobisphenol A polycarbonate, local chain motions in HFPAr should not be less than in PAr. If this assumption is correct, gas diffusivity in HFPAr would be expected to be much larger than in PAr because of the much smaller packing density of HFPAr, which is indeed the case as shown in Table VI. On the other hand, since tetrachlorobisphenol A polycarbonate exhibits a much more restricted local chain mobility than unsubstituted polycarbonate, one would expect TClPAr to behave similarly relative to PAr. As a result, even though TClPAr has a packing density smaller than that of PAr, the diffusivity of carbon dioxide in TClPAr is slightly less than that in PAr.

The larger gas solubility in PPha is related to its larger unrelaxed volume (therefore, larger Langmuir sorption capacity) and higher concentration of the polar ester group (therefore, larger Henry's law constant). In the case of HFPAr and TClPAr, the larger gas solubility may be attributed primarily to their lower packing and secondarily to their larger excess volumes. A more detailed discussion on the effects of unrelaxed volume on gas solubility can be found in a paper published earlier.[19]

An even more dramatic manifestation of the effects of packing density and local chain mobility on the diffusivity of gases in polymer is found in the third polymer groups.[20] In addition to PPha, which is synthesized by reacting phenolphthalein and a 50/50 mixture of iso- and tere-phthaloyl chlorides, PPha-tere and PPha-iso were prepared by reacting the bisphenol with pure terephthaloyl chloride and pure isophthaloyl chloride, respectively. Pertinent material data for these three polymers are reproduced in Table VII.

Table VII. Material Data for PPha-tere, PPha, and PPha-iso.

Polymer	Tg, °C	Density at 23°C, g/cm^3	δ, (cal/cm^3)$^{1/2}$
PPha-tere	299	1.297	11.0
PPha	279	1.298	11.0
PPha-iso	249	1.304	11.0

Since PPha-tere, PPha, and PPha-iso differ only in the relative amount of the two phthalate groups, mass density of the sample can be used to represent the packing density of the three polymers. Clearly, PPha-tere has the lowest packing density, and PPha-iso has the highest packing density although the difference is much smaller than for the two previous groups of polymers. Low temperature relaxation data for poly(ethylene isophthalate) suggest that local chain motions in poly(ethylene isophthalate) is more restricted than in poly(ethylene terephthalate). Extrapolated to the present case, it is natural to postulate that PPha-iso has less local chain motions than PPha-tere because of the meta bonding in PPha-iso.

As summarized in Table VIII, gas solubilities in these three polymers are essentially the same. Considering their similar chemical compositions and packing densities, one finds the solubility more or less as expected. In marked contrast, however, the diffusivities of both carbon dioxide and methane in PPha-tere are more than 100% larger than those in PPha-iso. Clearly, the differences in packing density and local chain motions discussed above must be the cause of this remarkable variation in gas diffusivity.

Table VIII. Permeability, Diffusivity, and Solubility Data for PPha-tere, PPha, and PPha-iso.

Polymer	P_{CO_2}	P_{CO_2}/P_{CH_4}	D_{CO_2} x 10^8	D_{CO_2}/D_{CH_4}	S_{CO_2}	S_{CO_2}/S_{CH_4}
PPha-tere	18.2	27	3.25	73.5	5.58	3.63
PPha	14.5	27	2.74	7.33	5.29	3.47
PPHa-iso	7.9	28	1.58	8.19	5.01	3.41

In conclusion, studies of modified polymers such as the two reported above do lead to helpful information on membrane design to increase both permeability and permselectivity. The results confirm the importance of reducing chain packing and torsional mobility to achieve these objectives. They completely confirm the conclusions of Koros et al.[11] from the elegant and extensive studies he and his colleagues have conducted over the years.

REFERENCES

1. T. Graham, <u>Phil. Mag.</u>,1866, <u>32</u>, 40.
2. A. S. Stern, *et. al.*, <u>Ind. Eng. Chem.</u>,1965, <u>54</u>, 49.
3. W. J. Koros, G. K. Fleming, S. M. Jordan, T. H. Kim and H. H. Hoehn, <u>Prog. Polym. Sci.</u>,1988, <u>13</u>, 339.
4. V. T. Stannett, <u>J. Membr. Sci.</u>,1978, <u>3</u>, 97.
5. A. G. Wonders and D. R. Paul, <u>J. Membr. Sci.</u>,1980, <u>5</u>, 63.
6. S. M. Jordan, W. J. Koros and G. K. Fleming, <u>J. Membr. Sci.</u>,1987, <u>30</u>, 191 and references contained therein.
7. M. D. Donohue, B. S. Mishas and S. Y. Lee, <u>J. Membr. Sci.</u>,1989, <u>42</u>, 3
8. R. T. Chern, W. J. Koros, H. B. Hopfenberg and V. T. Stannett, <u>ACS Symposium Series</u>,1985, <u>269</u>, 25.
9. S. M. Allen, M. Fujii, V. Stannett, H. B. Hopfenberg and J. L. Williams, <u>J. Membr. Sci.</u>,1977, <u>2</u>, 153.
10. V. T. Stannett and M. Szwarc, <u>J. Membr. Sci.</u>,1955, <u>16</u>, 89.
11. W. J. Koros and M. W. Hellums, <u>J. Membr. Sci.</u>,1989, in press,
12. R. T. Chern, F. R. Sheu, L. Jia, V. T. Stannett and H. B. Hopfenberg, <u>J. Membr. Sci.</u>,1987, <u>35</u>, 103.
13. R. T. Chern, L. Jia, S. Shimoda and H. B. Hopfenberg, <u>J. Membr. Sci.</u>,1989, in press,
14. A. Bondi, <u>J. Phys. Chem.</u>,1964, <u>68</u>(3),441.
15. W. J. Koros, private communication.
16. A. F. Yee and S. A. Smith, <u>Macromolecules</u>,1981, <u>14</u>, 54.
17. D. J. Massa and P. P. Rusanowsky, <u>Am. Chem. Soc., Div. Polym. Chem., Polym. Prepr.</u>,1976, <u>17</u>, 184.
18. K. L. Hoy, <u>J. Paint. Tech.</u>, 1970, <u>42</u>, 76.
19. F. R. Sheu and R. T. Chern, <u>J. Polym. Sci., Polym. Phys.</u>,1988, <u>26</u>, 883.
20. R. F. Sheu and R. T. Chern, <u>J. Polym. Sci., Polym. Phys. Ed.</u>,1989, <u>27</u>, 1121.

The Permselectivity of Polyorganosiloxanes Containing Specific Functionalities

A. J. Ashworth, B. J. Brisdon, R. England*, B. S. R. Reddy, and I. Zafar

SCHOOL OF CHEMISTRY AND *SCHOOL OF CHEMICAL ENGINEERING, UNIVERSITY OF BATH, BATH BA2 7AY, UK

1 INTRODUCTION

There is an expanding market for gas separation systems using membranes, with membranes already being used industrially to separate some gas mixtures (e.g. H_2/NH_3; CO_2/CH_4; O_2/N_2). Composite membranes used in such processes have a thin dense skin of a polymer providing the separation, coated over a mechanically strong and porous support. This has great potential in that it provides the freedom to tailor-make the polymer to satisfy the requirements of a particular separation, but that potential will only be fully realised when polymer films with the desired selectivity, permeability and durability become readily available.

Permeation through the polymer film is dependent on solution and diffusion of the gases in the polymer. The transport of a gas through such a membrane is therefore described by a permeability coefficient which can be expressed as the product of the diffusivity and the solubility coefficients of the gas in the membrane[1] as represented in eqn.(1).

$$P = D S \tag{1}$$

The permselectivity α_{AB} of a membrane for a binary gas mixture (A+B) can then be characterised by considering the ratio of the permeability coefficients of the two components in the membrane (P_A/P_B), when the pressure downstream from the membrane is much lower than their upstream pressure[1]. Moreover, using the relation in eqn.(1) the permselectivity can be partitioned in terms of the ratio of the diffusion coefficients, a diffusivity or mobility selectivity (D_A/D_B), and the ratio of the solubility coefficients, a solubility selectivity (S_A/S_B), according to eqn.(2).[2]

$$\alpha_{AB} = P_A/P_B = (D_A/D_B)(S_A/S_B) \qquad (2)$$

Diffusivity selectivity results from the ability of the polymer matrix to separate the components on the basis of size and shape, rather as a molecular sieve, while solubility selectivity is determined by the chemical and physical interactions between the permeants and the polymer.

The desired properties for a polymer for gas separa-tion membranes of high selectivity, high permeability, high mechanical and thermal stability are difficult to achieve. Polymers which have a high permeability usually have a low selectivity and vice versa. Therefore, there are two possible ways forward. Either the permea-bility of highly selective membranes must be improved, or the selectivity of high permeability membranes must be increased. Both approaches are being investigated with some success. Koros and co-workers[2] have adopted the first approach using modified glassy polymers, and we have chosen the second route, using rubbery polymers, which we believe are particularly promising, especially for difficult separations involving gas molecules of similar dimensions.

In order to achieve our aims, we selected poly-(dimethylsiloxane) as the basic substrate for our investigations. This polymer has a high permeability, is readily available, as are its precursors, and combines the advantages of the synthetic variability of organic polymers with the thermal and mechanical properties of inorganic polymers. Previous studies[3] have shown that substitution of increasingly bulkier functional groups in the side or backbone chains of silicone polymers results in a significant decrease in permeability for a given penetrant gas, due mainly to a decrease in diffusivity, while the solubility decreases to a much lesser extent. A selectivity enhancement can be achieved, but it is probably limited to mixtures of permeants that vary greatly in molecular size and shape, as when modified in this way, the polymer matrix then acts more like a "molecular sieve". Our research has centred on the production of highly permeable rubbery poly(dimethyl-siloxane) membranes containing side-chain functionalities which have a strong interaction with one of the components of a permeant gas mixture, thus providing a perm-selectivity based primarily on a difference in solubility, and hence not dependent upon size and shape differences in the gas molecules.

The removal of carbon dioxide from methane is of considerable current economic importance, and our first venture was to produce a range of polyorganosiloxanes with an ester functionality ($-COOCH_3$), reasoning that the ester group should provide a greater attraction for carbon dioxide than for methane, as carbon dioxide is known to interact with carbonyl groups present in ketonic and ester solvents[4]. The second functionality examined was that of the trifluoroester group ($-COOCF_3$), as one of the silicone polymers studied by Stern and co-workers[3], poly(trifluoro-propylmethylsiloxane), exhibited increased transport of carbon dioxide in preference to methane, due to the interaction between carbon dioxide and the fluoro group in the repeat side chain which increased the solubility selectivity. The third functionality investigated was that of a weakly basic amide group ($-NHCONH_2$), which would be expected to interact preferentially with carbon dioxide. This report therefore details the permeability and diffu-sion measurements over the temperature range of 35 to 100°C of carbon dioxide and methane in a series of poly-organosiloxane membranes functionalised to various degrees with ester, fluoroester and amide groups.

2 EXPERIMENTAL

Materials

The preparation of the functionalised membranes involved two stages. First, a known quantity of the organofunctionality was introduced into a linear poly-(methylhydridosiloxane), $Me_3SiO[MeSi(H)O]_n SiMe_3$, (where $n \sim 40$), by the platinum catalysed addition under anhydrous conditions of a 1-alkene, $CH_2 = CHCH_2R$, (R = $COOCH_3$, $COOCF_3$ or $NHCONH_2$). The synthesis and characterisation of typical examples of the resulting linear organo-

functional siloxanes $Me_3SiO[MeSi(H)O]_x\{MeSi[(CH_2)_3R]O\}_ySiMe_3$
have been reported elsewhere[5]. The lower molecular weight
polymers with high x:y ratios were gums and the higher
molecular weight polymers were organic solvent soluble
solids. Second, the formation of a membrane was achieved
by catalysed crosslinking of the remaining SiH units in
the linear organofunctional siloxanes with an
α,ω-dihydroxypoly(dimethylsiloxane), (disilanol), in the
presence of traces of tetramethoxysilane to ensure complete reaction. The reactants were mixed in the minimum
volume of toluene required to produce a clear fluid
mixture, in the presence of catalytic quantities of
dibutyltinlaurate. Crosslinking and solvent removal were
carried out at ambient temperature in a press using a 50
ton force, with the liquid polymer mixture held between
sheets of cellulose acetate in order to produce unfilled
membranes of 0.2 to 0.4 mm thickness. The thickness of
each membrane was determined by averaging many measurements with an electronic micrometer. The molecular weights
and the ratio of SiH groups to functionalised groups (x:y)
in the linear functionalised polymers, together with the
molecular weights of the disilanols, used in the preparation of a particular membrane are given in Table 1.
Unfilled membranes of pure poly(dimethylsiloxane), PDMS,
were prepared from both the high and low molecular weight
disilanols. Number average molecular weights were
determined at the Rubber and Plastics Research Association,
Shrewsbury by gel permeation chromatography based on
calibration with polystyrene using toluene as solvent.

Apparatus

    The permeabilities and diffusivities of the single
gases in these membranes were determined over the temperature range 35 to 100°C at the atmospheric pressure using
the dynamic method where the permeant is passed over one
face of the membrane and that permeating through the
membrane is collected by a carrier gas stream and
determined by a thermal conductivity detector. The method
developed by Pasternak et al.[6] has been used previously
by one of us[7] and the apparatus employed here was
identical except that a Pye 104 gas chromatograph thermal
conductivity detector replaced the flame ionization
detector. The detector was calibrated by using calibrant
gas mixtures of carbon dioxide and methane in helium
supplied by Electrochem Ltd., Stoke-on-Trent. The flowrates of the permeant gas and helium as carrier gas were
60 cm^3min^{-1}.

Table 1   Membrane Preparation from
$Me_3SiO[MeSi(H)O]_x\{MeSi[(CH_2)_3R]O\}_ySiMe_3$

Functionality Type	Mol% Si atoms	Linear Polymer x:y ratio	Mn	Disilanol Mn
ester	3.40	30:10	3680	71000
	5.48	20:20	4710	71000
	7.02	10:30	5630	71000
	8.35	10:30	5630	71000
	17.3	10:30	5630	1350
	21.4	10:30	5630	1350
Fluoroester	8.24	7:33	6360	71000
	15.0	7:33	6360	1350
	24.2	7:33	6360	1350
amide	8.71	8:32	5770	71000
	14.2	8:32	5770	1350

The permeability coefficient was determined from the steady state signal and the diffusion coefficient from the gradient of the signal before steady state is reached.[6,7]

The equilibrium sorption of CO_2 and CH_4 in some of the membranes was determined using a Sartorius Model 4102 electronic vacuum microbalance in conjunction with a Texas Instruments Bourdon pressure gauge allowing monitoring of weight and pressure changes to +0.01 mg and +0.01 mmHg respectively. The apparatus has been described previously.[8]

3 RESULTS AND DISCUSSION

The permeabilities and diffusivities determined for CO_2 and CH_4 in the ester functionalised membranes at 35°C and atmospheric pressure are shown in Figures 1 and 2. The permeabilities for both gases remain effectively constant, within experimental error, with increase in ester functionality, while the diffusivities for both gases show a decrease with ester functionality which becomes less marked as the functionality is increased. The results reported for a PDMS membrane with no ester functionality are for the membrane prepared from the high molecular weight disilanol, since unfilled PDMS membranes prepared from both the high and low molecular weight disilanols (Mn = 71000 and 1350 respectively) were found

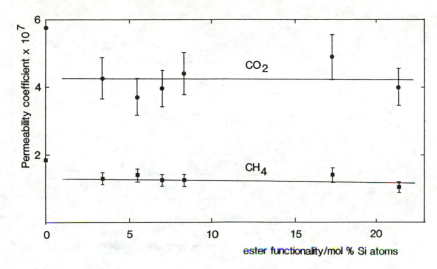

Figure 1   Permeability coefficients at 35°C
            [units = $cm^3(stp)cm/(cm^2 cmHg\ s)$]

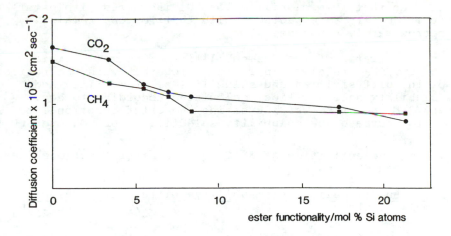

Figure 2   Diffusion coefficients at 35°C

to have very similar transport properties for the two
gases.   The solubilities of each gas in the membranes at
35°C estimated according to eqn.(1) from the permeabilities and diffusivities are reported in Figure 3.   It can

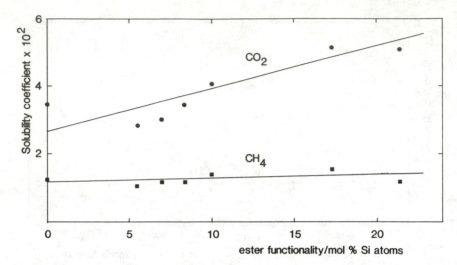

<u>Figure 3</u>   Solubility coefficients estimated from the
            transport measurements at 35°C
            [units = $cm^3(stp)/(cm^3 polymer)$]

be seen that the solubility for CO_2 increases with ester
functionality as expected, while that for CH_4 remains
approximately constant.

    The ratio of the permeabilities or permselectivity
for CO_2/CH_4 are given in Table 2, together with the ratios
of the diffusivities and solubilities, or mobility and
solubility selectivities, calculated according to eqn.1
and 2. The effect of increasing the ester functionality
is to increase the solubility selectivity for CO_2 over CH_4,

<u>Table 2</u> Selectivities at 35°C for the Ester Functionalised
         Membranes

Ester Functionality /mol% Si atoms	Permeability CO_2/CH_4	Diffusivity CO_2/CH_4	Solubility CO_2/CH_4
0	3.14	1.11	2.82
3.40	3.28	1.22	2.70
5.48	2.66	1.03	2.58
7.02	3.15	1.06	2.99
8.35	3.56	1.20	2.96
17.3	3.49	1.06	3.30
21.4	3.92	0.91	4.32

and as the diffusivity selectivity only slightly decreases, this results in an overall increase in the permselectivity. Addition of the ester groups in the polymer side chains has clearly increased the interaction with CO_2 in comparison with CH_4.

An independent check on the solubilities was made by determining the equilibrium sorption of CO_2 and CH_4 in some of the membranes at 30°C using a vacuum microbalance. The sorption isotherms determined were linear with the exception of that for the highest ester functionality of 21.4 mol% which is initially concave to the pressure axis. The isotherms determined for CO_2 and CH_4 with the

21.4 mol% functionalised membrane, together with the CO_2 isotherm for the 17.3 mol% membrane given for comparison, are shown in Figure 4. The solubilities determined and reported in Table 3 are the Henry's Law solubilities for the linear isotherms and the secant slope of the sorption isotherm[9] calculated at 1 atmosphere in the case of the 21.4 mol% functionalised membrane. The solubilities

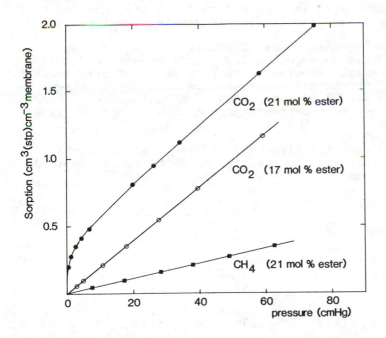

Figure 4 Sorption isotherms at 30°C

Table 3   Comparison of Gas Solubilities at 30°C for Ester
           Functionalised Membranes

Ester functionality /mol% Si atoms	Solubility x 10^3 /$cm^3(STP)/(cm^3 cmHg)^{-1}$		Solubility ratio CO_2/CH_4	
	$\underline{CO_2}$	$\underline{CH_4}$	balance	dynamic
0	17.5	6.25	2.80	2.82
7.0	17.9	5.77	3.10	2.99
17.3	18.1	5.56	3.26	3.30
21.4	24.3	5.47	4.44	4.32

determined from the quotient of the permeability and
diffusivity coefficients were a factor of two greater than
those determined directly and undoubtedly reflect the
assumptions made in their estimation by this method.
However, as can be seen in Table 3 the agreement between
the direct and indirect measurements of the CO_2 and CH_4
solubility ratio is good and confirms the increase in
solubility selectivity with increasing ester functionality.

    The permeabilities, diffusivities and derived
selectivities determined for the fluorester and amide
functionalised membranes are reported in Tables 4 and 5.
Results for the 8.24 mol% fluorester functionalised
membrane seemed promising, but, as can be seen in Table 5,
fluoroester membranes with higher mol% functionalities

Table 4   Permeability and Diffusion Coefficients at 35°C
           for the Fluoroester and Amide Functionalised
           Membranes

Functionality		Permeability^{a}x10^8		Diffusivity^{b}x10^6	
Type	Mol% Si atoms	CO_2	CH_4	CO_2	CH_4
fluoroester	8.24	41.3	11.7	9.0	8.5
	15.0	40.2	13.5	7.1	7.6
	24.2	43.6	15.5	8.1	9.2
amide	8.7	46.6	12.7	5.5	5.6
	14.2	43.1	11.2	7.0	6.8

{Units: ^{a} $cm^3(STP).cm/cm^2.cmHg.sec$ ^{b} cm^2/sec}

Table 5   Selectivities at 35°C for the Fluoroester and
          Amide Functionalised Membranes

Functionality		Permeability	Diffusivity	Solubility
Type	Mol% Si atoms	CO_2/CH_4	CO_2/CH_4	CO_2/CH_4
fluoroester	8.24	3.53	1.05	3.35
	15.0	2.98	0.93	3.20
	24.2	2.81	0.88	3.20
amide	8.7	3.67	0.99	3.71
	14.2	3.85	1.03	3.72

gave lower permselectivities and were inferior to the
ester functionalised membranes of comparable mol%
functionality. The 8.7 mol% amide functionalised membrane
appeared even more promising, showing a permselectivity
greater than that of the 8.35 mol% ester or 8.24 mol%
fluoroester functionalised membranes. The 14.2 mol% amide
membrane gave an even higher permselectivity, but unfortunately we were not able to carry out measurements
on unfilled and unsupported membranes with a greater amide
functionality than 14.2 mol% because of the brittle nature
of the resulting material.

Permeation and diffusion of the gases in all the
membranes gave linear Arrhenius plots over the temperature
range studied. The activation energies for permeation and
diffusion were determined from the measurements at a
minimum of four temperatures in the range 35 to 100°C. The
activation energies for permeation for the permeants did
not change greatly as the various functionalities were
increased, but that for CO_2 of ca.1 kJ/mol was much lower
that that for CH_4 of ca.8 kJ/mol. Thus permselectivity
for CO_2 over CH_4 resulting from the improved solubility
selectivity will become greater as the temperature
decreases.

Koros[9] has discussed the solubility selectivity of
membranes for CO_2/CH_4 separation and pointed out that the
solubility selectivities of polymers and chemically alike
low molecular weight solvents are similar. Thus the high
density of carbonyl groups in cellulose acetate imparts a
character similar to that of methyl acetate and results in
very similar relative solubilities for CO_2 to CH_4 of 11.4
and 11.5 respectively. It was also pointed out that there

was a relation between the solubility selectivity for CO_2 and CH_4 and the mass density of carbonyl groups for various polymers and low molecular weight solvents. The mass density of carbonyl groups was calculated simply by using the mass density of the various media and the weight fraction of the carbonyl functionality in the low molecular weight solvent or polymer repeat unit. The mass density of carbonyl groups in the 21.4 mol% ester functionalised membrane is 0.073 g cm^{-3}. The maximum mass density of carbonyl groups that could be obtained if a membrane were produced from a polymer with an ester group attached to each silicon atom would be ca. 0.18 g cm^{-3}. A higher mass density than that calculated for cellulose acetate (0.335 g cm^{-3}) could only be achieved by attaching two ester groups to each silicon atom. To date we have not been able to fabricate unfilled and unsupported membranes with an ester functionality greater than 21.4 mol% because of the brittle nature of the resulting material. However, it would appear that higher solubility selectivities of greater than 4.3 for CO_2/CH_4 in the 21.4 mol% ester membrane could be achieved if a polymer membrane with a higher degree of ester functionality could be produced, but that the maximum solubility selectivity for this gas pair would be about 11, similar to that found for cellulose acetate. A functionalised polyorgano-siloxane would have the advantage of greater permeability, poly(dimethylsiloxane) being 800 times more permeable than cellulose acetate.[2] The solubility selectivity found for the same gas pair in the amide functionalised polysiloxane membranes indicates that these are more promising than the ester functionalised membranes, but again attainment of a greater solubility selectivity than that determined will depend on the ability to introduce a greater functionality into the polymer.

ACKNOWLEDGEMENTS

The authors gratefully acknowledge the support of the Science and Engineering Council for this work.

REFERENCES

1. D.R. Paul and G. Morel, "Membrane Technology", in "Kirk-Othmer Encyclopedia of Chemical Technology", 3rd edn., Wiley, New York, 1981, Vol 15, p.92.
2. W.J. Koros, G.K. Fleming, S.M. Jordan, T.H. Kim and H.H. Hoehn, Prog.Polym.Sci., 1988, 13, 339.
3. S.A. Stern, V.M. Shah and B.J. Hardy, J.Polym.Sci., Polym.Phys.Ed., 1987, 25, 1263.
4. J.H. Hildebrand and R.L. Scott, "Solubility of Nonelectrolytes", 3rd edn., Reinhold Pub.Co., New York, 1950, p.249.
5. A.J. Ashworth, B.J. Brisdon, R. England, B.S.R. Reddy and I. Zafar, British Polymer J., in press.
6. R.A. Pasternak, J.F. Scimscheimer and J. Heller, J. Polym.Sci., Part A-2, 1970, 8, 467.
7. A.J. Ashworth and M.D. Wickham, J. Membrane Sci., 1987, 34, 225.
8. A.J. Ashworth and D.M. Hooker, "Progress in Vacuum Microbalance Techniques", ed. C. Eyraud and M. Escoubes, Heyden and Sons, London, 1975, Vol.3, p.330.
9. W.J. Koros, J.Polym.Sci., Polym.Phys.Ed., 1985, 23, 1611.

The Ultra-High Purity Challenge

Walter H. Whitlock

THE BOC GROUP, INC., *GROUP TECHNICAL CENTRE, 100 MOUNTAIN AVENUE, MURRAY HILL, NJ 07974, USA

ABSTRACT

Separation methods for the preparation of ultra-high purity (UHP) gases are examined. These gas separation methods include membranes, distillation, chemical conversion, and adsorption. In addition, the separation of particulate impurities is also examined. The suitability of these different separation methods for the preparation of UHP gases is investigated. Example applications of these techniques are discussed to illustrate their strengths and weaknesses.

The limitations in the current technology for manufacture and delivery of UHP gases are described. Areas of this technology are identified which would benefit from additional research and development work.

1 INTRODUCTION

During the latter half of the eighteenth century, Joseph Priestley and his contemporaries observed and documented many of the properties of the gases we know today. Joseph Priestley, himself, investigated the gases ammonia, carbon dioxide, carbon monoxide, nitrogen monox-

---

*The BOC Group Technical Center performs research and development work on behalf of Airco, BOC Ltd., Osaka Sanso Kogyo, and other group companies.

ide, nitrogen, hydrogen chloride, silicon tetrafluoride, and oxygen.[1] His published reports of this work describe many of the chemical characteristics of these gases, some of which he prepared for the first time.

It is probably impossible to know precisely what purities were obtained during these times. However, judging from the experimental technique and gas handling apparatus employed, impurity contents less than 1 part in a hundred or 10,000 parts per million would have been difficult to achieve. Gases collected over water in a pneumatic trough would have included approximately 2.5% water. Even for gases collected over mercury, air contamination from dead volumes, leaks, and permeation through flexible tubes made of leather and storage vessels made from animal bladder would probably have added impurities on the order of at least 1 percent.

For the work being done at the time, however, contamination at these levels was insignificant. The physical and chemical properties which were reported can be readily observed in gases containing 1% impurities. Also, during that era, chemistry was not sufficiently well understood that the effects of minor impurities, even if observed, would have been recognized for what they were.

Even so, commercial applications for these new gases began to be developed. Over time, as gases were put to more commercial uses, impurity levels below one part in 10,000 or 100 parts per million were found to be necessary. For example, during the last five years of the nineteenth century, commercial production of acetylene for lighting purposes was begun. By 1900, the impurities of phosphine, hydrogen sulfide, and ammonia had been identified as the source of undesirable haze accumulating in rooms with low levels of ventilation. The gas analytical techniques of the time were quite capable of detecting these impurities and it was recommended that they be limited to 0.01% in acetylene manufactured for lighting purposes.[2]

By the mid twentieth century, gas shielding applications in certain advanced metallurgical processes required impurities of less than 1 part per million. The manufacture and processing of refractory metals requires that oxygen and water impurity levels be below 1 ppm. This was also the time of the beginning of the semiconductor electronics manufacturing industry. When commercial production of silicon integrated circuits began in

1968, the semiconductor industry as a market for high
purity gases was established.

Today, for semiconductor manufacturing applications,
the need for high purity gases has been well established.[3]   Impurity levels in bulk argon and nitrogen
are typically requested to be below 0.01 ppm.   To what
extent this is actually necessary for successful manufacture is not precisely known.   However, impurity levels
below 0.3 ppm have been shown to be essential for achieving superior properties in the materials which constitute
today's advanced products.

The link between high performance in advanced
products and extremely low levels of impurities in raw
materials was probably first apparent to the semiconductor industry. However, as new products have been developed in other industries, new applications for ultra-high
purity materials have been recognized. This can be seen
in the history of fiber optics development for long
distance communication, the original concept of which is
generally credited to Koa and Hockham.[4]   In 1966 they
discussed the requirements for a dielectric fiber wave-
guide to be used for long distance optical communications.   They reported a measured attenuation of <200
dB/km for a bulk sample of commercial fused quartz
compared to 4000 dB/km for a typical optical glass. They
predicted that attenuation below 20 dB/km could be
obtained by a combination of improved fiber packaging and
purification of the bulk glass to reduce certain impurity
ions below 1 ppm.

By the late 1970's, processes were developed for the
manufacture of optical fiber preforms from the high
purity raw materials employed for semiconductor manufacture. The internal chemical vapor deposition process was
developed to convert high purity gas phase reagents into
solids on the inner wall of a heated reaction tube.
Using this technique, solid particles of precisely
controlled composition are generated by thermal decomposition and deposited onto the inner tube wall. The
preform is finished when the tube has been filled with
glassy layers formed by these particles. The preform is
then heated to fuse the particles and drawn into continuous lengths of optical fiber hundreds of kilometers
long.   By using high purity gases as reagents and improved fiber manufacturing processes, attenuations of 1
dB/km have been achieved.[5]   While material purity was
not the only factor responsible for the decrease in
attenuation, high purity materials are absolutely re-

quired. The availability of ultra-high purity reagent gases has contributed to the establishment of an entirely new communications technology in the brief span of 20 years.

In other manufacturing industries, such as advanced ceramics and surface coatings, one can also see a trend towards employing higher purity materials. There is no reason to doubt that the need for ultra-high purity materials will emerge in other industries as well. Finding the means to economically manufacture and deliver large quantities of ever higher purity gases is the "Ultra-High Purity Challenge".

2 ULTRA-HIGH PURITY GASES

The purity requirements for certain materials used in advanced semiconductor manufacture have reached the point where impurity levels in the starting materials of 1 part in 10^{12} have been related to reduced yield and device performance. For example, the resistivity required in the epitaxial layer deposited from silane on a silicon wafer may exceed 10,000 Ohm centimeter. This demands that trace dopant contamination in the silane be held to less than one part in 10^{12}. There is no universally accepted definition of what impurity levels can be called ultra-high purity. However, gases with impurities specified below 0.1 parts per million must be manufactured and handled differently if that specification is to be maintained. Requiring that certain impurities be maintained below 0.1 ppm is thus a good working definition of ultra-high purity.

A discussion of ultra-high purity gases is incomplete without addressing particulate impurities and particulate control. Particulates are very significant impurities in virtually all applications of ultra-high purity gases. In the semiconductor industry, gas specifications requiring less than 10 particles per cubic foot greater than 0.1 micron diameter are typical. This level is only tolerated as a specification because current particle measurement instrumentation is insensitive to lower levels.

3 COMPARISON OF GAS SEPARATION METHODS

Many factors affect the suitability of a separations process for a particular application. This is evidenced by the fact that so many different separation methods and combinations of these methods are found to be suitable in actual practice. However, for the manufacture of ultra-high purity gases, there is at least one important characteristic, and that is the ability to achieve a high degree or extent of separation. Of course, any process can in principle be run many times or staged to obtain the required degree of separation. However, a process which can achieve ultra-high purities in fewer stages will be at an advantage. With this single criterion in mind, gas purification techniques using membranes, liquid vapor equilibrium, chemical conversion, adsorption, and particle filtration will be examined. The objective is to look for fundamental limitations in their application to the preparation of ultra-high purity gases. Of course, by examining only the fundamentals, many of the practical limitations in using these techniques will be ignored. But it is always conceivable that these non-fundamental limitations will be circumvented by new and ingenious developments in the future.

The degree or extent of purification achieved by different techniques for separating an impurity component from a bulk gas component is commonly expressed in terms of a separation factor. It can be defined by the following ratio of concentrations obtained for a gas passing from the inlet to the outlet of a single separation stage.

$$\text{Separation factor} = SF = \frac{(C_i \; / \; C_b)_{outlet}}{(C_i \; / \; C_b)_{inlet}}$$

where: subscript i = impurity component
       subscript b = bulk component

This ratio will be used to compare the degree of separation which can be achieved by different gas separation methods.

4 PURIFICATION BY MEMBRANE PROCESSES

Separation by membrane processes between bulk and impurity components occurs due to their different rates of transport through the membrane. A driving force is required to move these components through the membrane and usually this is a concentration difference resulting from an applied pressure difference. The property of a membrane describing the rate of transport is permeability. Table 1 lists some permeability data for selected gases and membrane materials.

Table 1

Membrane Permeability Data – $cm^3/cm^2/s$

Gas	Material Polycarbonate[6] 298 K	80%Pd–20%Ag[7] 623 K
H_2	0.00087	2.5
O_2	0.000093	<2.5(-10)
N_2	0.000018	<2.5(-10)

For membrane separations, the separation factor can usually be approximated by a ratio of permeabilities. Separation factors for selected gases and membrane materials are shown in Table 2.

Table 2

Purification by Membranes
Calculated Separation Factors

Composition Impurity/Bulk	Membrane Material Polycarbonate 298 K	80%Pd–20%Ag 623 K
H_2 / N_2	47	>1(10)
H_2 / O_2	9.3	>1(10)
O_2 / N_2	5	–
N_2 / O_2	0.2	–
O_2 / H_2	0.1	<1(-10)
N_2 / H_2	0.02	<1(-10)

The single stage separation factors for the polycarbonate membrane are not sufficiently low that ultra-high purities can be reached without staging. This performance is also typical for other polymeric membrane materials and as a consequence, most single stage membrane processes are not useful for ultra-high purity gas separations. There are, however, ways of staging the membrane process such that high separation factors can be obtained in a continuous membrane column.[8] Unfortunately, the high energy cost per unit of pure product produced by this staging technique makes it uneconomical.

However, the palladium/silver membrane does have a single stage separation factor that is very favorable for

the removal of oxygen and nitrogen from hydrogen. In
practice, the membrane cost is low enough that single
stage purification is commonly used for the preparation
of ultra-high purity hydrogen. Hydrogen purifiers based
on purification by a palladium based membrane are commercially available.[9]

   Until recently, residual impurities remaining in
hydrogen purified by this technique were thought to
result mainly from small holes in the membrane which
allowed small amounts of unpurified hydrogen to leak
through. However, a different contamination mechanism
has been reported recently.[10] Methane contamination was
found to result from the reaction of hydrogen and a small
carbon impurity in or on the palladium/silver membrane.
Under typical membrane operating conditions, this mechanism accounted for as much as 0.05 ppm CH_4 in the ultra-
high purity product.

   For hydrogen purification by membrane, the calculated separation factor shows that the gas separation
process can be performed satisfactorily. However,
mechanisms of contamination unrelated to the purification
method can take away much of the gain in purity.

5 PURIFICATION BY LIQUID VAPOR EQUILIBRIUM PROCESSES

Purification by liquid vapor equilibrium occurs because
of a volatility difference between the bulk and impurity
components. As a result, a change in phase from either
liquid to vapor or vapor to liquid is accompanied by a
change in composition. The necessary driving force to
cause the phase change is usually transport of heat.

   The degree of purification achieved with one stage
of liquid vapor equilibrium can be expressed as a separation factor computed from the impurity component k-
factor. The k-factor for a particular component is
defined as its mole fraction in the vapor phase divided
by its mole fraction in the liquid phase under conditions
of equilibrium.

$$k = \frac{(\ C_i/(C_i+C_b)\ )_{vapor}}{(\ C_i/(C_i+C_b)\ )_{liquid}} \longrightarrow \frac{(\ C_i/C_b\ )_{vapor}}{(\ C_i/C_b\ )_{liquid}}$$

Under high purity conditions, the impurity concentration
is very low relative to the bulk concentration. This
allows the C_i term to be dropped when summed with C_b. For

a phase change from liquid to vapor, the separation factor thus becomes numerically identical to the component k-factor.

Table 3

Purification by Liquid Vapor Equilibrium
Calculated Separation Factors at 1 atm.

Composition Impurity/Bulk	Liquid --> Vapor Temperature, K	k-factor
H_2 / O_2	90.3	>1000
N_2 / O_2	90.3	3.9
H_2 / N_2	77.5	>400
O_2 / N_2	77.5	0.25
N_2 / H_2	20.5	5.9(-7)
O_2 / H_2	20.5	2.6(-8)

Table 3 lists some separation factors for a liquid to vapor phase change based on calculated k-factor data for selected gases. The calculated separation factors are very favorable for the removal of oxygen and nitrogen impurities from hydrogen. Because of this, a single stage of liquid-vapor equilibrium could be employed for the manufacture of ultra-pure hydrogen from a more conventional grade.

The data show a relatively unremarkable separation factor for the removal of oxygen from nitrogen. Fortunately, the amount of heat needed for driving the phase change does not vary greatly with respect to composition. This balance in heat requirement between separation stages has allowed very efficient multiple staging methods to be developed. Experience shows that multiple stages of liquid vapor equilibrium can be applied to gas separation with high efficiency. As shown in Table 4, cryogenic distillation is very effective at reducing the level of contaminants found in air to manufacture ultra-high purity nitrogen. Overall separation factors (computed from the impurity and bulk concentration at the inlet and outlet of a multiple stage separation process) of less than $5.0(10)^{-7}$ have been achieved for certain impurities in bulk nitrogen by purification systems employing multiple stage cryogenic distillation.

Table 4

Manufacture and Purification of Bulk Nitrogen
Overall Separation Factors

Impurity	Product ppm	Feed Air ppm	Separation Factor
H_2	<0.1–1.0	0.5	<0.1–2.0
CO	0.1–0.3	10.0	0.01–0.3
O_2	0.2–0.5	210000	9.5(-7)
CO_2	<0.1	300	<2.8(-4)
H_2O	<0.1	10000	<1.0(-5)

Today, cryogenic distillation of air is the only method used commercially for the manufacture of large quantities of ultra-high purity nitrogen. Calculated separation factors and actual distillation experience show that the gas separation process can be performed satisfactorily. However, from the eventual user's point of view, it is not sufficient simply to manufacture an ultra-pure gas, it must also be delivered to the point of use. As will be discussed later, mechanisms of contamination unrelated to the purification method can take away much of the purity.

6 PURIFICATION BY CHEMICAL CONVERSION PROCESSES

Purification processes based on chemical conversion utilize one or more chemical reactions which convert the impurity or bulk component (or both) into a different chemical compound. The properties of the resulting compounds (reaction products) make it easier to separate the bulk and impurity components. The driving force which causes the chemical reaction to occur is a reduction of the free energy of the total system. This free energy must be supplied by outside means to convert the reaction products back to their original form.

In principle, either the impurity component or the bulk component can be the object of chemical conversion. The original Brin process for the manufacture of commercial oxygen from air relied on the conversion of BaO to BaO_2, with the subsequent release of oxygen when BaO_2 was converted back to BaO. However, because of the large amount of free energy which must be cycled back and forth between reactants and products, small inefficiencies can result in large losses. As a result, gas separations based on chemical conversion of the bulk component are seldom economical. However, for converting the small quantities of material associated with low levels of impurities, gas purification by chemical conversion can work very well.

Chemical conversion is quite practical for the removal of trace levels of certain reactive impurities from nitrogen or other inert gases. Trace oxygen can be chemically converted by the reaction

$$2\ Cu\ +\ 1/2\ O_2\ -->\ Cu_2O$$

carried out in a fixed bed. The oxygen impurity is converted to a solid compound, which separates from the remaining bulk gas. Likewise, trace hydrogen can be chemically converted to water by the reaction

$$H_2\ +\ 1/2\ O_2\ -->\ H_2O$$

which can take place either in the gas phase or on the surface of a suitable catalyst. The product water can be separated from the bulk gas more effectively by other methods than was the case for the original hydrogen impurity.

Chemical thermodynamics (with some generous assumptions) can be used to estimate the separation factors which could be achieved under the most favorable circumstances. If a state of chemical equilibrium is assumed to exist in a system consisting of gaseous reactants and single phase solid components, then the equilibrium concentrations of the reactants can be calculated by standard techniques.

Table 5

Purification by Chemical Conversion
Calculated Separation Factors at 1 atm.

System (Copper) Impurity/Bulk	Reaction Product(s)	Fixed Bed of Catalyst 1300 K	700 K	300 K
H_2 / O_2	H_2O g	8.4(-8)	2.6(-16)	1.7(-40)
O_2 / H_2	H_2O g	2.9(-20)	2.7(-37)	1.1(-85)
H_2 / N_2	NH_3 g	1.0	1.0	0.57
N_2 / H_2	NH_3 g	0.94	4.1(-2)	9.0(-12)
O_2 / N_2	Cu_2O s	0.62	2.9(-12)	4.2(-46)
N_2 / O_2	$N_2O+NO+NO_2$ g	0.61	1.0	1.0

Table 5 lists some calculated separation factors obtained by chemical conversion in a bed of finely divided copper metal. The gas phase concentrations were calculated under conditions of chemical equilibrium resulting from an initial 1.0 ppm impurity concentration. The tabulated separation factor is the ratio of the resulting impurity concentration to the initial 1 ppm concentration. The copper can participate in the chemical reactions in two ways: it can serve as a catalyst and it can be consumed as a reagent.

The data in Table 5 illustrate several features of gas purification by chemical conversion. Trace hydrogen in oxygen and trace oxygen in hydrogen are converted to water and, provided the water is removed efficiently by other means, very low impurity concentrations can be obtained. The major uncertainty is whether or not the reaction kinetics are fast enough. There are certainly better catalysts than copper metal for the reaction of hydrogen and oxygen. Higher temperatures can be used to speed up the kinetics so long as the separation factor remains favorable.

The same cannot be said for the removal of trace hydrogen from nitrogen and nitrogen from hydrogen by a hypothetical conversion to ammonia. While the separation factor for removal of nitrogen from hydrogen at 300 K appears useful, the reaction kinetics at this temperature are so slow (with even the best catalyst) that effectively no chemical conversion takes place. Raising the temperature does not help much since unfavorable separation factors will result from shifts in equilibrium.

The removal of trace oxygen from nitrogen using finely divided copper works very well in practice. The oxygen reacts with the copper to form Cu_2O, which forms a separate solid phase with a very favorable separation factor. This reaction is known to occur rapidly enough to be used at 300 K for gas purification.

Trying to remove trace nitrogen from oxygen by conversion to nitrogen oxides is a losing proposition. Since the separation factor is unfavorable, the slow reaction kinetics do not even need to be considered. And, if the situation were not already bad enough, the finely divided copper metal has been completely converted to CuO by the bulk oxygen.

For the most part, chemical conversion is very attractive for the manufacture of ultra-high purity gases from more conventional grades of high purity gases. As the examples have illustrated, it is preferable to operate at temperatures near ambient, both to avoid the cost of heating and cooling large quantities of gas and also to obtain more favorable separation factors. Provided a way can be found to achieve the necessary kinetics for the desired reactions, low temperatures can also serve to inhibit undesired side reactions, such as that already mentioned in connection with hydrogen purification by palladium membrane.

Some very effective purification schemes have been developed for the small scale purification of reactive gases used in the manufacture of semiconductor devices. Porous organometallic polymers are available which react instantly and irreversibly with water, oxygen, carbon dioxide, chlorosilanes, chlorofluorocarbons, arsine, phosphine, and to a lesser but still useful degree, carbon monoxide.[11]

For gas purification by chemical conversion, calculated separation factors and actual purifier experience show that the gas separation process is performed satisfactorily. However, as will be seen later, mechanisms of contamination unrelated to the purification method can take away much of the gain in purity.

7 PURIFICATION BY ADSORPTION PROCESSES

Gas purification by adsorption, stated simply, requires that the impurity component be adsorbed to a relatively higher concentration in the adsorbed phase than in the gas phase. This tendency for preferential adsorption of the impurity component can be expressed as a separation factor. However, calculating separation factors for realistic situations is significantly more complex for adsorption than the previously discussed separations processes. Even to generate some separation factors for illustrative purposes requires substantial calculation.

In applying adsorption to gas purification, transport phenomena are important to a greater degree than most other separations processes. Separation factors based on binary adsorption equilibrium, while important, do not always provide enough information to predict a successful separations process. The bulk separation of nitrogen and oxygen by adsorption on carbon molecular sieves is a good example. Based on an equilibrium separation factor analysis, there should be no separation whereas, in fact, because of differing rates of adsorption, nitrogen is the product. Nevertheless, most purification done by adsorption is based on equilibrium behavior.

Many adsorption equilibrium models have been derived for single component adsorption. Some of these have been modified to predict binary adsorption isotherms, one example being the potential theory of adsorption.[12] According to this adsorption model, under conditions of sufficiently low concentration, such as when a trace impurity is adsorbed from a bulk component, adsorption equilibrium will simplify into a single term proportional to a power of the partial pressure of the impurity in the gas phase. Based on this analysis, the equilibrium separation factor obtained at low impurity concentrations can be extrapolated to even lower concentrations.

Analysis of pure component adsorption using statistical mechanics has shown that all adsorption isotherms should approach the Henry's law relation at sufficiently low component pressures.[13] Under these conditions the quantity adsorbed is directly proportional to partial pressure.

Table 6

Purification by Adsorption on Active Carbon
Calculated Separation Factors

Composition Impurity/Bulk	Temperature K	Separation Factor
O_2 / N_2	123	0.6
N_2 / H_2	77	1.0(-5)
O_2 / H_2	90	1.7(-4)

Some typical gas phase separation factors are shown in Table 6. They were obtained by inverting the calculated ratio of concentrations of the impurity to bulk component in the adsorbed state. In reality, the single stage gas phase separation factor is a function of the amount of adsorbent present and the change in gas phase concentrations must be found by mass balance. However, provided that sufficient adsorbent is available and used efficiently, the gas phase concentration ratio can be made to approach the inverted adsorbed phase concentration ratio.

To make efficient use of a bed of adsorbent, the isotherm should have a convex shape with respect to the ordinate, called type I or favorable adsorption behavior. Extrapolation of binary isotherm models shows that if favorable adsorption behavior exists at a sufficiently low impurity concentration, it will continue to exist at even lower concentrations. Thus, one has good reasons to expect that purification by adsorption can, for certain favorable systems, achieve very low impurity levels.

In practice, applying adsorption to a gas purification process is more an empirical process than a predictive one. Although predicting the behavior of gas purification by adsorption is complex, gas purification by adsorption techniques has been successfully applied in situations where ultra-high purities are required. For example, the liquefaction of industrial quantities of helium and hydrogen on a continuous basis could not be accomplished without reducing impurities to levels well below 0.1 ppm. This purification is accomplished by adsorption in beds of silica gel and active carbon at temperatures near 77K. In Japan, where liquid hydrogen is not available for commercial use in the electronics industry, hydrogen (and helium) purifiers based on low temperature adsorption are commercially available with

outlet specifications below 0.01 ppm for a broad range of
impurities.[14]

However, as has been the case with other gas separation methods, even though high purities can be obtained,
mechanisms of contamination unrelated to the purification
method can take away much of the gain in purity.

8 PURIFICATION BY FILTRATION PROCESSES

For gas purification by filtration, as the gas
passes through the separation stage, particles are left
behind on the filter medium. Filters remove particles by
three basic mechanisms; inertial impaction, interception,
and diffusion, which are described using the single fiber
filter model shown in Figure 1. These three mechanisms
identify the different methods of particle transport from
the gas to the filter medium, in this case a single
fiber, where once contact is made the particle adheres
tightly.

Particle collection by inertial impact occurs for
large diameter particles with sufficient momentum which
do not follow the abrupt changes in direction of gas
streamlines near the fiber. These particles tend to
remain on their original path and impact the fiber
surface.

Medium diameter particles with less momentum will
follow gas streamlines. Collection by interception
occurs if the streamline passes within one particle
radius from the fiber. The particle will intercept the
surface and remain collected.

Certain streamlines exist, however, that will not
result in particle interception. Brownian motion of very
small diameter particles causes them to deviate randomly
from the gas streamlines. This motion significantly
increases the probability of their contacting the fiber,
even if originally they were traveling on a non-intercep-
ting streamline. This particle diffusion mechanism
dominates for particle diameters less than 0.1 micron.

The particle removal efficiency achieved using
filtration can be found by summing the contributions to
particle removal from these three filtration mechanisms.
For particles composed of the same material, the minimum
removal efficiency usually occurs at diameters between
0.1 and 0.3 micron. These particles are too large for

Figure 1

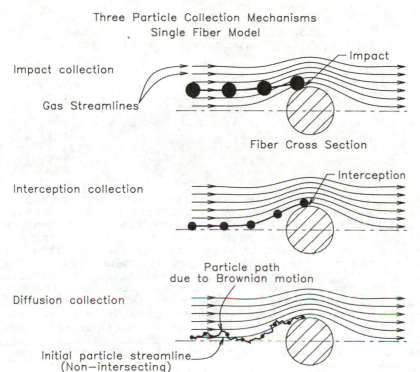

Three Particle Collection Mechanisms
Single Fiber Model

Impact collection

Gas Streamlines

Impact

Fiber Cross Section

Interception collection

Interception

Particle path
due to Brownian motion

Diffusion collection

Initial particle streamline
(Non—intersecting)

diffusion and too small for interception to be very effective.

Today, high efficiency filters are made from a large area membrane with numerous small diameter pores. By challenging these filters with generated sodium chloride particles, separation factors can been computed from upstream and downstream particle density measurements.[15] Typical results are shown in Table 7. The most penetrating particle size for a filter consisting of 0.2 micron diameter pores is calculated to be between 0.035 and 0.089 micron. Thus, these separation factors represent the worst case performance; the separation factor will be substantially better for particles of larger and smaller diameter. For all practical purposes, particle separation by filtration can be said to achieve complete removal of particles from the gas stream.

Table 7

Purification by Filtration
Calculated Separation Factors at 1 atm.
PVDF Membrane Containing 0.2 Micron Pores

Particle Diameter, Microns	Concentration, cm^{-3} Inlet	Outlet	Separation Factor
0.089	5(6)	<1.5(-2)	<3(-9)
0.067	5(6)	<1.5(-2)	<3(-9)
0.046	5(6)	<1.5(-2)	<3(-9)
0.035	5(6)	<1.5(-2)	<3(-9)

This is not to say that particulate contamination can be completely eliminated by filtration. Recent tests have shown that, while high quality filters can achieve particulate contamination levels at or below present day detection levels, particulate contamination shed from the filter and released from the piping downstream of the filter is a major source of contamination.[16] Gaseous contamination is also known to be released by filters.[17] Typical impurities include water, oxygen, carbon dioxide, and nitrogen. These impurities are thought to arise from two sources: impurity components which have been adsorbed on the internal surfaces of the filter and impurity components which have permeated within the filter components.

Desorption from the internal surfaces of the filter is responsible for the rapid outgassing of atmospheric gases which were adsorbed on the component surfaces. Since high efficiency filter media is made with numerous small pores, high surface areas are present. Also, the filter media is often made of gas permeable materials with significant gas solubility, such as polyvinylidene fluoride (PVDF). Impurity components can migrate into the filter medium during exposure to air and then migrate slowly back out of the medium to contaminate a high purity gas.

For gas purification by filtration, the separation factors show that the gas separation process can be performed satisfactorily. However, as will be seen in more detail later, mechanisms of contamination unrelated to the purification method can take away much of the gain in purity.

9 FACTORS LIMITING THE PURITY OF UHP GASES

For the separation technologies based on membranes, distillation, chemical conversion, adsorption, and filtration, examination of separation factors has shown that each method has some applicability to the preparation of ultra-high purity gases. Under the proper conditions and for a suitable gas and impurity combination, each separation method can achieve ultra-high purities. If one particular separation method is insufficient, another method can be found with better performance. If no single method is sufficient, more than one method can be combined.

Combining separation methods can be a very effective way to prepare ultra-high purity gases. A process has been developed for the manufacture of ultra-high purity inert gases from lesser purity grades. It functions at ambient temperature and can remove oxygen, hydrogen, carbon monoxide, carbon dioxide and moisture to below 0.010 ppm on the scale of several hundred cubic meters per hour.[18] This method combines a chemical conversion method for removal of oxygen, a physical adsorption method for removal of water and carbon dioxide, and a chemisorption method for removal of hydrogen and carbon monoxide in the absence of stoichiometric oxygen.

So, in practice, a set of very effective separation methods is available to achieve ultra-high purities. Because science and engineering have been employed to solve problems arising in the separation of gases, the ability to perform gas separations and manufacture high purity gases has been well developed. Based on an examination of the separation methods alone, there should be no problem in attaining ultra-high purities.

However, experience has shown that obtaining ultra-high purities is difficult. Some of the examples used to illustrate the separation methods have also shown some of the problems. To supply ultra-high purity gases, they must first be manufactured with sufficient purity and then packaged and delivered without degrading this purity. Improving the packaging and delivery part of the supply problem has largely been carried out on an "as needed basis". The limitations in achieving ultra-high purities today arise not so much from insufficient separation but from contamination added during packaging and distribution operations. No matter how pure the gas

is as manufactured, its purity will not be maintained at
the point of use if contaminants are picked up in passing
through the delivery system.   For ultra-high purity
gases, it is the performance of the gas distribution
system, not the original purification method, which
largely determines gas purity at the point of use.

The gas delivery system can have very detrimental
effects on the purity of the gas which passes through it.
The effects of contamination derived from air through
external leakage are obvious. Moisture contamination can
also arise from desorption from the surfaces of piping,
fittings, and vessels.   Particle contamination can be
released from imperfections in the piping and dirt left
over from manufacturing and fabrication operations. For
all forms of contamination, impurities released from dead
ends can significantly degrade gas purity.

Particles, aside from being a form of contamination
themselves, can also contribute to outgassing. Particles
of hydrated or carbonated salts, such as calcium sulfate
and calcium carbonate found in numerous building materials, will contribute contamination while chemically
converting to their anhydrous and oxide forms. They can
also serve as adsorbents, slowly releasing gaseous
contaminants into the high purity gas stream.

Attaining ultra-high purity gases today requires a
suitable gas distribution system.[19]   To achieve an ultra-
high purity distribution system, there must be   an
absence of dead zones, an absence of external leakage, an
absence of outgassing, and an absence of particulate
contamination.[20] Dead zones are obvious sources of carry-
over contamination and must be designed out of the
system. External leakage can arise from the diffusion of
air through small holes, which takes place despite a
reverse pressure gradient.   Leakage can also occur
because of permeation, especially through organic polymers used for gasketed seals.   Outgassing is a general
term applied to the release of any gaseous contaminants
from the materials of construction.   It can result from
impurities adsorbed on surfaces and dissolved within the
bulk material of both system components and particulate
contamination. As we have seen for hydrogen purification
by membranes, it can also result from the generation of
gases by chemical conversion.

Over the years, a substantial body of technology has
been developed for minimizing the re-contamination of
high purity gases by the distribution system. A complete

description of this technology will not be attempted. However, the use of high cleanliness components with low outgassing surfaces is extremely important. Low permeability materials, typically 316L stainless steel, must be employed throughout the system. Minimum dead volume designs and leak tight fabrication techniques are essential. High efficiency filtration, using low permeability materials to reduce outgassing from dissolved contaminants, is employed frequently.

As the need for purer and purer gases continues, keeping the gas pure once it enters the distribution system will become an even more difficult task. Finding better methods for the distribution of ultra-high purity gases is an intense area of current research and development.[21] Special surface treatments are being developed to reduce outgassing of adsorbed contaminants from piping and components.[22] As a result of work in this area, advances in the performance of gas delivery systems and the resulting ultra-high purity gases have led to advanced materials properties.[23]

10 MEETING THE ULTRA-HIGH PURITY CHALLENGE

As we anticipate future needs for even higher purity gases and think about how to achieve them, certain observations can be made based on our examination of the preceding separation methods. These observations suggest areas where limitations in current technology could be improved by applying research and development effort.

We have seen the importance of minimizing contamination added by the distribution system after the gas has been purified. One significant contamination mechanism is desorption of impurities from the piping surfaces. While work is being done on surface treatments to minimize adsorption, I believe this is an area where further progress can be made. So far as I know, no examination of surface treatments has been made on a fundamental level with the objective of minimizing their adsorption capacity. There is a significant collection of knowledge and expertise in adsorption fundamentals within the discipline of gas separations. Surfaces with low adsorption capacity could be developed and applied to the piping and components used for the distribution of ultra-high purity gases.

We have also seen the effects of elevated temperature. Increased permeability and chemical reactions,

both of which release unwanted impurities, have resulted.
Thus, if separation processes and gas distribution can be
accomplished at reduced temperatures, there is less
likelihood for recontamination. As we have seen, the
separation factors for chemical conversion are sensitive
to temperature. While this often favors lower temperatures, reduced reaction kinetics can make chemical
conversion impractical. Further effort could be profitably employed in the development of reduced temperature
chemical conversion methods. Since many of the chemical
methods require reactive surfaces, advances in the
technology for manufacturing high surface area catalysts
could be applied.

Finally, despite all efforts to manufacture a pure
gas and minimize its contamination during distribution,
some recontamination will occur. The problem of recontamination during distribution was addressed in high
purity water systems many years ago. Current practice is
to use both a primary and secondary purification system.
A diagram illustrating this concept is shown in Figure 2.

The primary system processes feed water and supplies
purified water to a final continuous circulation loop.
The loop distributes the water to the points of use.
Built into this circulation loop is a secondary purification system to continuously repurify and circulate the
water in the loop. This purification scheme removes
trace impurities caused by interaction of the pure water
with its container, the piping system. No one would ever
consider using bottled water for high purity applications; it simply cannot be maintained at high purity
without continuous repurification.

Recently, this concept has been employed to supply
high purity chemicals used in semiconductor manufacture.[24]
This recirculation concept can also be applied to ultra-
high purity gases. As in Figure 2, a primary supply of
purified gas is sent to a secondary purifier. The secondary purifier both purifies and circulates the gas in a
loop past the points of use. Use of this concept would
supply the highest purity product at the point of use.

In Joseph Priestley's time, the apparatus used for
gas handling was responsible for much of the contamination.   This is also the case today for the highest
purity gases, even though impurity contents have been
lowered by a factor of 10^6 or more in the intervening 200
years.   If the need for lower impurity levels in gases
continues to progress the way it has in the past, it is

Figure 2

Re—circulation and Re—purification process

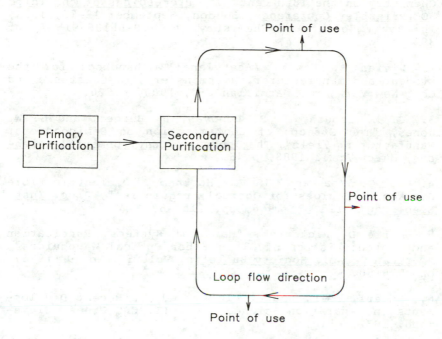

interesting to speculate about the way our present day gas distribution technology will compare with that of 200 years in the future. It will probably seem as primitive then as the leather and animal bladder of Joseph Priestley's era seem to us now.

REFERENCES

1. R. G. W. Anderson, "From Chaos to Gas: Pneumatic Chemistry in the 18th Century", <u>Proceedings of the Third BOC Priestley Conference</u>, London, September 12-15, 1983, The Royal Society of Chemistry, ISBN 0-85186-915-7, p. 395.

2. Vivian B. Lewes, 'Acetylene: a handbook for the student and manufacturer, Westminster', A. Constable and Co.; New York, the Macmillan Co., 1900, p.520.

3. C. M. Osburn, R. P. Donovan, H. Berger, and G. W. Jones, "The Effects of Contamination on Semiconductor Manufacturing Yield", <u>The Journal of Environmental Sciences</u>, March/April 1988, p.45.

4. K. C. Koa and G. A. Hockham, "Dielectric-Fibre Surface Waveguides for Optical Frequencies", <u>Proc. Inst. Elec. Eng.</u>, 113 [7] 1966, pp. 1151-58.

5. Merle D. Rigterink, "Material Systems, Fabrication and Characteristics of Glass Fiber Optical Waveguides", <u>American Ceramic Society Bulletin</u>, Vol. <u>55</u>, No. 9 (1976), pp. 775-80.

6. Rogers, Fels, and Li, in Li (ed.), 'Recent Developments in Separation Science', vol II, CRC Press, Cleveland, 1972, p.107.

7. A. G. Knapton, "Palladium Alloys for Hydrogen Diffusion Membranes: A Review of High Permeability Materials", <u>Platinum Met. Rev.</u> vol. <u>21</u>, (1977), pp. 44-50.

8. Hwang, S. T., and Thorman, J. M., "The Continuous Membrane Column", <u>A.I.Ch.E Journal</u>, 26(4), 558 (1980).

9. Johnson Matthey, P.O. Box 733, Valley Forge, PA.

10. K. Sugiyama, F. Nakahara, M. Abe, T. Okumura, T. Ohmi, and J. Murota, "Detection of Sub ppb Impurities in Gases Using Atmospheric Pressure Ionization Mass Spectrometry", <u>9th ICCCS Proceedings</u>, 1988.

11. Steven J. Hardwick, Raymond G. Lorenz, and Daniel K. Weber, "Ensuring Gas Purity at the Point-of-Use", <u>Solid State Technology</u>, October 1988, pp. 93-96.

12. S. D. Mehta and R. P. Danner, "An Improved Potential Theory Method for Predicting Gas-Mixture Adsorption Equilibria", <u>Industrial and Engineering Chemistry Fundamentals</u>, V. <u>24</u>, N. 3, p. 325-330 (1985).

13. T. L. Hill, <u>J. Chem. Phys.</u>, V. <u>18</u>, 1950 p. 246.

14. Osaka Sanso Kogyo Ltd., 1-14, Miyahara 4-chome, Yodogawa-ku, Osaka 532, Japan.

15. M. A. Accomazzo, K. L. Rubow, B. Y. H. Liu, "Ultrahigh Efficiency Membrane Filters for Semiconductor Process Gases", <u>Solid State Technology</u>, March 1984, p. 141.

16. Mark L. Malczewski and John R. Sarratori, "Evaluating Point-of-Use Filtration in Process-gas Contamination Control", <u>Microcontamination</u>, August 1988, p. 43.

17. B. Gotlinsky, S. Tousi, E. Edlund, and W. Murphy, "Outgassing of Inorganic Filters", <u>Microelectronic Manufacturing and Testing</u>, May 1989, p. 19.

18. W. R. Weltmer and W. H. Whitlock. "Purifier Considerations for Ultra High Purity Inert Gases", <u>Microelectronic Manufacturing and Testing</u>, October and December 1987, p. 34.

19. Jeffrey Eagles, Neil Downie, and Masatoshi Goto, "Delivery Systems for ULSI-Grade Gases to Ensure Purity at the Point of Use", <u>Microelectronic Manufacturing and Testing</u>, V. <u>12</u> No. 2, February 1989, p. 33.

20. Yoichi Kanno and Tadahiro Ohmi, "Components Key to Developing Contamination Free Gas Supply", <u>Microcontamination</u>, December 1988, p. 23.

21. David A. Toy, "Gas Purity and Wafer Fabrication: Is the Tail Wagging the Dog?", <u>Semiconductor International</u>, April 1989, p. 114.

22. K. Sugiyama, T. Ohmi, T. Okumura, and F. Nakahara, "Electropolished, Moisture-Free Piping Surface Essential for Ultrapure Gas System, <u>Microcontamination</u>, January 1989, p. 37.

23. Satoshi Mizogami, Yutaka Kunimoto, and Tadahiro Ohmi, "Ultra Clean Gas Transport from Manufacturer to Users by Newly-Developed Tank Lorries and Gas Storage

Tanks", <u>Proceedings of the 9th Int. Symp. on Contamination Control</u>, September 1988, p. 352.

24. John B. Davison, Joe G Hoffman, and Wallace I. Yuan, "Ultrapurification of semiconductor process liquids", <u>Proceedings of the 9th International Symposium on Contamination Control</u>, Los Angeles, California, September 26-30, 1988, pp 442-445.

The Industrial Practice of Adsorption

G. K. Pearce

KALSEP (BP), WEY HOUSE, LANGLEY QUAY, WATERSIDE DRIVE, LANGLEY, BERKSHIRE
SL3 6EX, UK

1 INTRODUCTION

Current Status of Industrial Adsorption Processes

The separation of gases by adsorption can offer a low cost alternative to separation technologies such as adsorption or cryogenic distillation. In addition, adsorption can provide a route for carrying out separations that have previously been considered difficult.

The main advantage of adsorption in comparison with other separation techniques is that high recoveries of the adsorbed component can be achieved yielding a non-adsorbed stream of very high purity.

The principal applications for adsorption separation processes currently are relatively few, namely H_2 purification, air separation, CO_2 removal, natural gas sweetening and drying, and emmission control. For these applications, adsorption processes are low energy and have a competitive capital cost.

Most industrial adsorption processes use standard adsorbents which are commodity materials. These adsorbents often have relatively low selectivities. Whilst they are capable of achieving high purity of the non-adsorbed component, the adsorbed component is often obtained as a low concentration by-product stream. This either then forms a low value waste stream or has to be recycled or reprocessed at a relatively high cost. Low adsorbent selectivity has therefore restricted the range of applications for adsorption processes.

BP's Experience in Adsorption Processes. BP's experience in adsorption processes for gas separation is both
as a supplier and as a user of pressure swing adsorption
(PSA). Through its Kaldair subsidiary, BP is one of the
world's major suppliers of N_2 PSA plants, using the
Bergbau Forschung process based on a special adsorption
carbon. Kaldair has supplied PSA plants for many applications ranging from blanketing and purging on offshore oil
production platforms to controlled atmosphere storage for
fruit.

     BP is also a major user of PSA. For example in its
chemical and oil operations, at Hull and Grangemouth
respectively, BP uses PSA for H_2 purification. At Hull,
the H_2 stream contains 2% v/v CO from a Cold Box and
needs to be purified so that it can be used for hydrogenation. At Grangemouth, hydrocarbon contaminants are
removed from the H_2 stream also for a hydrogenation application. In both cases the contaminant is obtained as a
low concentration by-product (< 10% v/v) in H_2, and is
fed to the fuel system.

     Development in Industrial Adsorption Practice. At
present, adsorption processes in the chemical and oil
industries tend to be used in conjunction with other separation processes. If adsorption processes are to play an
increasing role both in existing areas of application and
in new applications, it will be necessary to develop
adsorbents with improved selectivity. High selectivity
adsorbents will reduce the need for recycling and reproces-
sing, thereby reducing energy and capital cost; also the
possibility of achieving more 'difficult' separations will
be opened up.

     Recognising this, BP initiated a research programme
in the early 1980's with the target of developing high
selectivity adsorbents. This paper discusses in detail
one particular development by BP Chemicals in which a
selective adsorption process was developed for CO/N_2
separation. The process is in an advanced stage of development but has not yet been commercialised.

     2 DISCUSSION

Background to the CO Recovery Target

     BP Chemicals has a strong commercial interest in
carbonylation processes to produce acetic acid and acetic

anhydride, and it therefore has an interest in CO separation and CO recovery technologies.

Several technologies are commercially available for separating CO from components such as H_2 and CO_2. However, CO/N_2 is a more difficult separation, and BP Chemicals identified this separation as an important future application within its operations.

A research programme was therefore initiated in 1983 to develop a selective adsorbent which would allow CO to be recovered from CO/N_2 by PSA with high efficiency. It was anticipated that other applications within BP Chemicals such as H_2 purification of H_2/CO and ethylene recovery from polyolefin plant purge gas streams might also be achievable through the same development.

Development of Cu(I)Y Zeolite. A large number of modified adsorbents were screened as potential candidates for selectively adsorbing CO from CO/N_2 mixtures. Zeolite Y was the preferred starting material since it has low N_2 physisorption, it can be readily ion exchanged, and it has received relatively little attention in the patent literature.

A range of transition metal cations were screened which would give CO selectivity through a weak chemisorption effect. Of these, Cu(I) was selected for further development; patents were granted in 1987 covering the use of Cu(I)Y for CO and ethylene recovery.

The preparation of Cu(I)Y involves an initial exchange step of Cu(II)Y $(NO_3)_2$ with NaY, followed by drying and then reduction at $350° - 450°$ C with CO or H_2/N_2.

Preliminary Evaluation of the Cu(I)Y PSA Process
Laboratory evaluation of Cu(I)Y showed that CO could be removed from CO/N_2 mixtures using a PSA cycle with a sub-atmospheric pressure regeneration step. The high selectivity for CO resulted from reversible chemisorption. This enabled both the non-sorbed N_2 stream to be obtained at high purity as in conventional PSA, together with a high CO concentration adsorbed product.

The CO and N_2 adsorption isotherms in Figures 1 and 2 show the classical chemisorption and physisorption

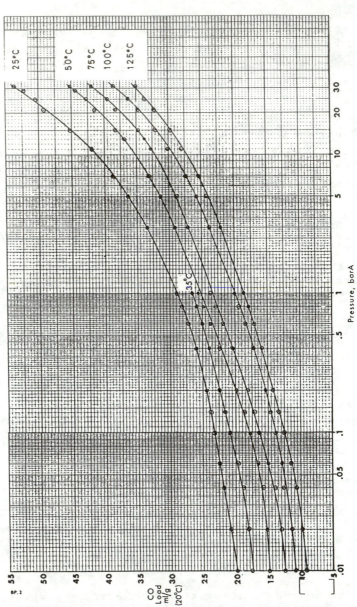

CO Adsorption vs Pressure   Cu(I) Y Zeolite

Figure 1

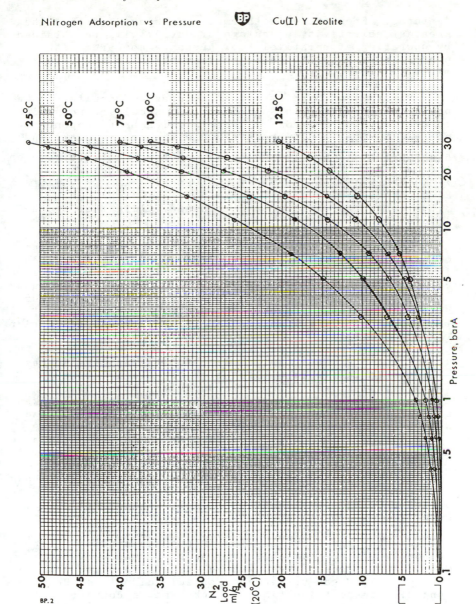

Nitrogen Adsorption vs Pressure   **BP**   Cu(I) Y Zeolite

Figure 2

features. From these isotherms, it is evident that the
greatest selectivity advantage of Cu(I)Y will be obtained
by operating at relatively low pressure compared with
conventional PSA, since N_2 physisorption is then minimised.

   Pilot Plant Studies. The encouraging results of the
preliminary evaluation showed that it should be possible
to develop a Cu(I)Y PSA process in which both components
of a CO/N_2 feed could be obtained with high recovery and

high purity. A pilot plant was therefore built to develop
the process. The objectives of the study were:
1. to characterise adsorbent performance
2. to provide process design data
3. to conduct a chemical and mechanical life test on the
   adsorbent.

   The tower of the pilot plant simulated a streamline
of a full scale process plant. Design details are given
in Figure 3 and a flowsheet in Figure 4. The available
cycle options for the study, listed in Figure 5, included
co and counter current flow, and product purge to ensure
high purity of the adsorbed CO product. The purpose of
the product purge was to displace N_2 reject and feed prior
to regeneration.

   A 4 bed process was proposed for the application as
illustrated in Figure 6, in which the single tower of
the pilot plant went through each of the steps of the
cycle sequentially. Where steps between beds were linked
as in equilibration, or repressurisation, bottled gases
were used to complete the simulation. The use of gas
between beds is dependent not only on composition but also
on pressure.

   Details of the zeolite charge for the study are shown
in Figure 7. The pilot plant study was targeted at a
specific CO/N_2 opportunity anticipated at the Hull site,
details of which are given in Figure 8.

   Pilot Plant Results. The effect was studied of the
following process variables on the performance of the
Cu(I)Y PSA process: Feed gas velocity, Feed gas concentration, Feed gas pressure, Particle diameter, Bed
length, Regeneration time, co-current vs counter
current, Cycle time.

   A key process design parameter is mass transfer zone
length (MTZL). Figure 9 shows the dependence of MTZL

Cu(I)Y P.S.A. Pilot Plant
DESIGN

* SINGLE TOWER
        -height 4.5m
        -diameter 0.05m
        -plant streamline

* COUNTER-CURRENT OPERATION

* SEQUENTIAL CYCLES
        -repressurisation
        -loading
        -depressurisation
        -vacuum regeneration

* AUTOMATIC CONTROL
        -Gould Modicon 84 P.L.C.
        -solenoid operated valves

* STREAM FLOWRATE MEASUREMENT
        -rotameter
        -collection

* STREAM COMPOSITION ANALYSIS
        -gas chromatography
        -mass spectroscopy
        -infrared spectroscopy

Figure 3

Cu(I)Y P.S.A. Pilot Plant
COUNTER-CURRENT FLOWSHEET

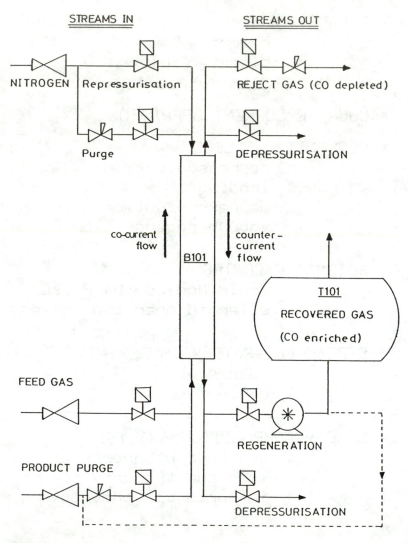

Figure 4

Cu(I)Y PSA Pilot Plant

CYCLE OPTIONS

* REPRESSURISATION
    - Counter-current with N_2
      (Equalisation Step)
    - Co-current with Feed

* DEPRESSURISATION
    - Co-current (Equalisation Step)
    - Counter-current

* PURGE BY STRONG ADSORPTIVE
        (Co-current)

* VACUUM REGENERATION (Counter-current)
    - N_2 purge assisted

Figure 5

Cu(I)Y P.S.A. Process
PROPOSED 4 BED CYCLE

<u>Bed no</u>

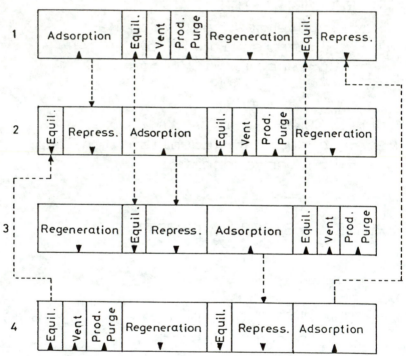

▲ Co-current

▼ Counter-current

Cycle time = 12 minutes

Figure 6

Cu(I)Y P.S.A. Pilot Plant

SECOND CHARGE CHARACTERISTICS

* SPHERE DIAMETER $=2-3mm$

* CROSS SECTIONAL AREA $=0.195dm^2$

* BED LENGTH (L) $=4.512m$

* CHARGE WEIGHT $=4.589kg$

* BULK DENSITY $=0.521kg/dm^3$

* TOTAL VOIDAGE $=82\%$

Figure 7

Cu(I)Y PSA Pilot Plant

REVISED CO RECOVERY APPLICATION

* FEED GAS $-77\%V/V$ CO $(1437kg/h)$
                $23\%V/V$ N_2

* SUPPLY PRESSURE   2.5 bar A

* TEMPERATURE   ambient

* SUPERFICIAL 0.10 m/s   (L=4.5m
  VELOCITY   3:1 Aspect ratio)

$$(U_1/U_2) = (MTZL_1/MTZL_2)^{0.81}$$

For $U_s = 0.1$ m/s MTZL = 1.83 m
                     LUB = 0.50 m

Figure 8

EFFECTS OF SUPERFICIAL VELOCITY ON
MASS TRANSFER ZONE LENGTH FOR
2-3mm SPHERES

PSA PILOT PLANT

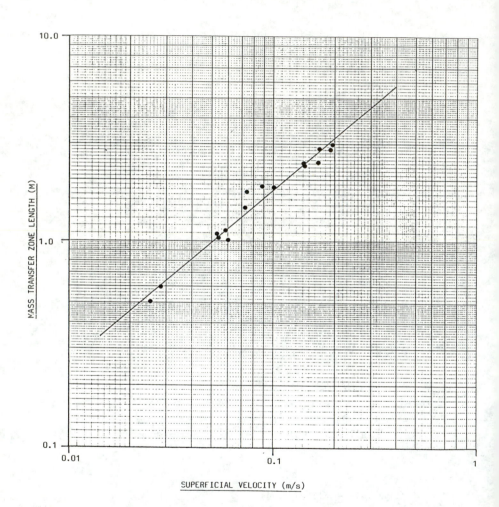

Figure 9

on velocity, and Figure 10, the relationship between
bed utilisation and velocity, for 2-3 mm spheres. MTZL
is shorter and bed utilisation greater for 1 mm spheres
cf 2-3 mm spheres at a given velocity, but this is at
the expense of considerably reduced regeneration efficiency. This is shown for co current flow in Figure 11
for different particle diameters and bed lengths. The
bed with 2-3 mm spheres benefited considerable from low
pressure drop, particularly under conditions of low
pressure regeneration. 2-3 mm spheres were therefore
selected for further process optimisation.

Counter current flow improves recovery dramatically
as expected, giving a N_2 reject with <200 ppm CO after
only 180 seconds of regeneration. The final pressure in
the regeneration step was 100 mbar A.

In most cases cycles were terminated prematurely.
This allowed a first stage co-current depressurisation to
remove pure N_2 above atmospheric pressure without breakthrough to the CO adsorption front.

Figure 12 shows the mass balance obtained from the
straightforward 4 bed cycle illustrated in Figure 6. A
projected mass balance is also shown for a cycle in which
the product purge, which is responsible for virtually
the entire CO loss, is recovered in a co-current step
prior to counter current N_2 repressurisation. The cycle
modification could be achieved by reducing the regeneration time from 180 seconds to 145 seconds, probably
resulting in a very slight rise in CO in the N_2 reject.
The remaining 35 seconds of the cycle could then be used
to recover CO from the product purge giving an estimated
overall CO recovery >99%.

It may then be possible to improve the CO product
purity of 96% by lengthening the product purge cycle at
the expense of a slightly increased overall cycle time
(and hence plant cost).

The possibility of separating both components of the
CO/N_2 binary giving high purity and high recovery for both
product streams therefore appears to be achievable in a
simple efficient cycle involving a 4 tower process.
During the course of extensive pilot plant studies, no
chemical deactivation of the Cu(I)Y zeolite was measured.
Mechanical integrity of the adsorbent also appeared to be
good with no measurable increase in pressure drop as a
result of fines formation.

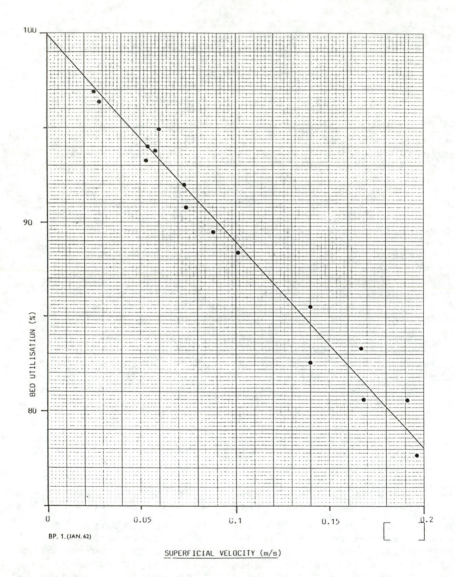

PSA PILOT PLANT   EFFECT OF SUPERFICIAL VELOCITY ON
                                                       BED UTILISATION FOR 2-3 mm SPHERES

BP. 1. (JAN. 62)

SUPERFICIAL VELOCITY (m/s)

Figure 10

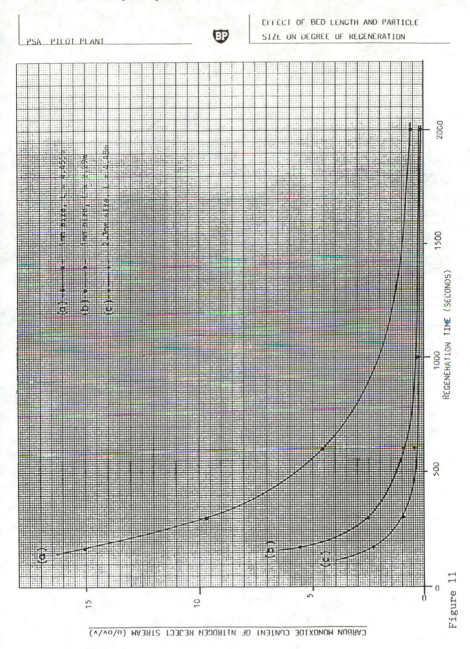

PSA PILOT PLANT

BP

EFFECT OF BED LENGTH AND PARTICLE
SIZE ON DEGREE OF REGENERATION

Figure 11

Cu(I)Y PSA PILOT PLANT

		CO	N2
FEED	%v/v	74.7	25.3
	ndm3/cycle	30.0	10.1
CO PRODUCT	%v/v	96.0	4.0
	ndm3/cycle	26.1	1.1
PRODUCT PURGE	%v/v	37.6	62.4
	ndm3/cycle	3.8	6.3
N2 REJECT	%v/v	0	100
	ndm3/cycle	0	2.8

KALSEP Sept 1989

Figure 12

Process Economics. A process economic comparison has
been carried out for three potential applications in BP
Chemicals. In each case significant benefits (i.e 2-3
year paybacks) have been demonstrated where the recovered
components have chemical (rather than fuel) value.
Benefits have also been shown in using the selective
Cu(I)Y PSA process rather than conventional separation
technology in achieving these separations.

3 CONCLUSIONS

Adsorption processes are widely used by industry for
gaseous separation. However, the range of applications
has been relatively limited due to low adsorbent selectivities. Advances have relied more on process innovation
than adsorbent development. BP has experience of this
type of adsorption process both as a supplier of N_2 PSA

plants through Kaldair Gas Services, and as a user of
for example, PSA for H_2 purification in its chemical and

oil operations.

    If adsorption processes are to make a greater impact
in the future, tailored adsorbents will be required with
improved selectivity. These highly selective adsorbents
will not only increase the number of applications for
adsorption in current markets, but will also open up new
areas.

    BP Chemicals development of Cu(I)Y PSA has demonst-
rated that CO and ethylene recovery can be an attractive
processing option. For example, both components of the
CO/N_2 binary could be obtained at high purity and high

recovery in a conventional 4 tower PSA cycle.

The Origins of Photoelectrochemistry

Mary D. Archer

NEWNHAM COLLEGE, CAMBRIDGE CB3 9DF, UK

1 INTRODUCTION

Photoelectrochemistry is a hybrid word embracing hybrid and diverse phenomena resulting from the absorption of light (usually in the ultra violet or visible range) by electrochemical systems. The action of the light must change the voltage or current to produce a photovoltage or a photocurrent.

Excepting the small thermal effects that are almost universal following light absorption by electrochemical systems, photoelectrochemical phenomena arise from the movement of charge - either electrons or ions - across the illuminated electrode/electrolyte interface. This in turn happens because electronically excited states have been created following absorption of light by the electrode, by a pigment coating on the electrode, or by the electrolyte.

Substantial photoelectrochemical effects are obtained only when the electronically excited states created by light absorption are rather long-lived. For example, although it is possible to detect the photoemission of electrons into solution from clean irradiated mercury electrodes, this occurs only in very low quantum yield because excited hole-electron pairs created by light absorption in metals relax extremely rapidly, within a femtosecond or so, before any charge has time to move across the electrode/electrolyte interface. Similarly, the absorption of light in solution may produce some photo-chemistry resulting in the changed concentration of some species to which the electrode responds. However, ions and molecules diffuse slowly in solution and effects arising from absorption of light in solution, sometimes called *photogalvanic effects,* are consequently also small.

Semiconductor electrodes produce by far the most significant photoelectrochemical effects. These are in essence *photovoltaic effects* because the absorption of photons of energy sufficient to excite electrons from the valence band of the semiconductor to its conduction band creates long-lived excited states which are also highly mobile; excess electrons and holes in a semiconductor can be highly mobile, and in an electric field they also migrate in opposite directions. The space-charge layer of a semiconductor is thus a fertile breeding ground of substantial photovoltaic effects. These have been intensively investigated since the early seventies as a possible means of converting solar energy into electrical energy or into stored chemical energy in the form of an electrolytically produced fuel such as hydrogen. In the course of these investigations, a new and sometimes stormy alliance has been forged between electrochemistry and solid-state physics, for the band theory of semiconductors has proved to be the key to understanding semiconductor photoelectrochemistry.

It is the pre-history of that seventies explosion of interest into photoelectrochemistry that I want to trace. Detailed accounts of more recent developments are provided by Myamlin and Pleskov[1], Pleskov and Gurevich[2] and Finklea[3].

2 BECQUEREL and BIOT

All photoelectrochemists with a sense of history acknowledge Edmond Becquerel as the father of their field. (He was also, in the literal sense, the father of Henri Becquerel, the discoverer of the emanation of uranium.) It was Edmond's work, extending over several decades, that laid the foundations both of photoelectrochemistry and photovoltaics. The announcement of his discovery was followed by a vigorous nineteenth-century controversy.

On 30 July 1839 Edmond's father, Antoine Cesar Becquerel, Professor of Physics at the Musee d'Histoire Naturelle in Paris, communicated to the Academie a brief note[4] by Edmond which described a novel electrochemical method of estimating the rate of certain photochemical reactions. *"On manque de procedes physique pour reconnaitre l'action de deux substances, l'une sur l'autre, sous l'influence de la lumiere,"* Edmond Becquerel had noted, "One lacks a physical process for following the

reaction of two substances in the presence of light," and
he described an electrochemical arrangement which he
was of the opinion did just that. He immersed two platinum
electrodes, one in methanol or ether and the other in
aqueous ferric perchlorate, and observed a small current to
flow between them when the interface between the two phases
was irradiated - over and above, that is, the small
transient current that flows on first contact on the dark.
What happens, we would say, is that as the photoredox
reaction occurs at the interface, a slight excess of
negative charge builds up on the aqueous side of the
interface and a corresponding slight positive charge on the
organic side, and the resulting liquid junction potential
causes a small current to flow.

Becquerel experimented with *des ecrans de diverse
nature* - filters of various colours - and, in reporting his
observations, assumed proportionality between the reaction
rate, the observed photocurrent and the *"nombre des rayons
chimiques"*. [5]   This, he remarked, rendered obsolete the
earlier method of Madame de Sommerville and M. Biot of
estimating the chemical effect of various kinds of light
from the tint produced in a sheet of AgCl-coated paper.

Six days after Edmond's first communication had been
read, it drew a speedy response[6] from the M. Biot in
question, Jean Baptiste Biot, Professor of Physics in the
College de France and a long-established authority on
optical dispersion and polarisation.   Correctly noting
that the observation of a physical effect does not
necessarily provide a linear measure of its cause, Biot
commented that Becquerel had failed to establish that the
chemical effect of the light was measured by the magnitude
of the electric current passing under the action of light,
and moreover that various colours of light could produce
different chemical effects.

Becquerel was silent for a few months and then in
November 1839 he communicated, again through his father, a
longer paper[7] containing the first observations of a
photoelectrochemical effect on coated electrodes.   Could,
he asked, the two electrodes themselves, immersed in a
single homogeneous medium, produce a photoelectrochemical
effect?   He found that a small photocurrent was produced
by the use of two clean platinum electrodes of different
temperatures, but that a much stronger effect was obtained
if one electrode had previously been heated to redness and
was then illuminated.   The question as to how the light was
acting in this case was one, he remarked, *"a laquelle il
est difficile de repondre,"* but he did postulate an answer,

attributing the effect to a thin layer of 'corpuscles' on
the surface of the heated electrode.[8]   Becquerel then
deliberately coated brass and platinum electrodes with
various substances such as halogens and silver halides and
observed some substantial photogalvanic effects. Again, he
asserted in conclusion that *"des effets electriques ...
peuvent servir a determiner le nombre des rayons chimiques
actifs."*

Again, Biot dissented the following week.[9]   *"Je
croirais inutile de repeter [mes objections] ici;
mais ...."* - but he did of course repeat his objections. A
fortnight later, Edmond was back in print defending his
hypothesis with vigour.[10]   After Becquerel's father had
read his son's Note, Biot again remarked that he could not
accept that the magnitude of the current was a measure of
the chemical effect of the light, and a week later he too
was back in print[11] with a fine polemic against Becquerel
and his supposed actinometer. *"The Academy will readily see
that what follows has a scientific purpose. I have nothing
personally against a young man whose zeal and inventive
mind I appreciate ..."* and away he went, citing the colour
sensitivity of the Daguerre photographic process, which had
been developed two years earlier, in support of his view
that the magnitude of the photoelectrochemical effect was
not a question of the number of rays but of their colour,
and the finding of the young Victor Regnault that an effect
persisted in Becquerel's original two-phase system after
the sun had gone down on it, when the number of incident
rays must surely have diminished.

Undaunted, Becquerel continued his work on electrodes
coated with silver halides[12], devising a 'galvanic photo-
meter' consisting of two pure silver plates coated with
silver iodide immersed in acidified water and connected to
a sensitive galvanometer.[13]   The controversy as to what
produced the photoeffects rumbled on for years, hardly
capable of resolution before the quantum theory of light
and electricity provided a natural link between one
absorbed photon and the one electron it caused to move.

3 EARLY PHOTOGALVANIC WORK

Meanwhile, interest in photoelectrochemical effects
obtained with dissolved coloured substances and light-
insensitive electrodes continued. In the 1890's,
Rigollet[14] observed photoelectrochemical effects from
dissolved fluorescent substances, and it was discovered
that irradiation of certain dye-containing solutions
seemingly altered their electrical conductance[15,16].

Hodge[17] showed that this was due to the creation of a
potential difference between electrodes immersed in the
light and the dark regions of the solution. Goldmann[18]
ascribed such photopotentials to the ejection of electrons
by the light-absorbing molecules. However, Baur[19],
Schiller[20], Trümpler[21] and others concluded that
the effects were due to photochemical reactions,
particularly electron-transfer reactions. Swensson[22]
established that photopotentials could be produced by
illumination of the electrolyte only. He concluded that
light reversibly changed a light-absorbing species A in
solution to another species B at a rate proportional to the
concentration of A and the irradiance, changing the 'energy
content' of the solutions and producing a corresponding
change in the electrode potential. Swensson's work
provides the first clear recognition of the nature of the
photogalvanic effect, in which the absorbed photon energy
is temporarily stored as the Gibbs free energy of the
product species. Accordingly, Copeland *et al.*,[23], in
their comprehensive review of early PEC work, used the term
Swensson-Becquerel effect to denote the PG effect.

Swensson's findings were confirmed by Grumbach[24] and
Murdock[25] using flow and other techniques to illuminate
the electrolyte but not the electrodes. Theoretical work
by Rule[26] and Ghosh[27] followed in which photopotentials
were attributed to the creation of a concentration cell by
the photoredox reactions of the electroactive dye.
Russell[28] explained photopotential growth and decay in
terms of diffusion.

Most early workers used solutions containing a dye and
a sacrificial electron donor. An early report by Rideal &
Williams[29] on the thermally-reversible Fe(II)/I_2 system
was non-committal as to the nature of the electrode
processes. However, Sasaki & Nakamura[30] realized that
the 'photopotential' in this system depended on the
concentration of iodine, whether produced by irradiation or
added in the dark. Holst[31,32] studied the (slightly
exergonic) dark and photochemical reduction of methylene
blue by various substituted phenylhydrazines, and
correlated reaction rates with the Gibbs energy for
reaction, assessed from the redox potentials of the couples
involved.

Eugene Rabinowitch, later distinguished for his work
on photosynthesis, introduced the term *photogalvanic
effect*, defining it as 'the change in the electrode
potential of a galvanic system, produced by illumination
and traceable to a photochemical process in the body of the

electrolyte.'[33]   From work on the Fe(II)/thionine system, he recognized that light-induced shifts in redox equilibria produced the photogalvanic effect and that photopotentials were determined by the composition of the photostationary state.

Interest during the fifties and sixties in novel power sources for military and space applications prompted reviews by Rowlette[34] and Kerr[35] which were pessimistic as to the prospects for photogalvanic cells.   More fundamental reviews were written during this period by Sancier[36] and Kuwana[37].

4 EARLY SOLID STATE WORK

Very substantial progress on the theoretical treatment of photogalvanic cells had thus been achieved by the 1960's.   It was at about the same time that semiconductor photoelectrochemistry assumed its modern form, though as a result of the entirely separate development of the theory of solid state semiconductors, to which we should therefore briefly turn.

The first observation of photoconductivity, by Smith[38] in solid Se, led to the discovery of the solid-state PV effect by Adams & Day[39], who constructed a photocell by sealing platinum contacts to the ends of a Se rod and noted that the effect could form the basis of a light meter.   Later, Fritts[40] constructed the first Se PV cell, consisting of crystalline Se, a thin front-surface metal grid and an iron base.   Similar Se photocells are still widely used as light meters.

The Se photocell is a barrier cell, in which the photovoltage is developed across the interface between two dissimilar materials, one of which is usually a metal.   Up until the early 1940's, barrier cells were the only photovoltaic cells investigated.   Two other barrier cells, namely the Cu/Cu_2O cell and the thallous sulfide cell, were developed during the early 1900's.   The photosensitivity of oxidized copper surfaces had first been noted by Hallwachs in 1904[41], but over 20 years passed before a Cu/Cu_2O barrier PV cell was constructed.[42]   A similar barrier layer was discovered at oxidized surfaces of thallous sulfide[43].   In spite of considerable commercial use of these photocells as light meters, none of them achieved solar power conversion efficiencies greater than ~1%.

Early in 1941, R.S. Ohl, a metallurgist at Bell
Telephone Laboratories, made a very important discovery.
On crystallization of a melt of commercial "high purity"
silicon, a "well-defined barrier having a high degree of
photovoltaic response" was detected. The barrier was
formed from the unequal distribution of impurities as the
Si crystal grew from the melt. This accidental discovery
marks the beginning of work on *p–n* junction devices.
Although Ohl patented his device[44], he did not publish
his work in the general literature for over ten years[45];
hence the initial discovery did not generate much interest
until the early 1950's.

The development of efficient solar cells was greatly
aided by the formulation of *p–n* homojunction theory by
Shockley[46] in 1949. This led to the fabrication by
Fuller & Ditzenberger[47] of the first diffused *p–n*
homojunction and, shortly thereafter, in 1954, to the
announcement by Chapin, Fuller and Pearson[48] of Bell
Telephone Laboratories of a diffused *p–n* homojunction Si
cell with a solar power-conversion efficiency of ~6%. The
first commercial Si solar cells were developed specifically
for applications in space satellites and were quite
expensive. Nevertheless, the explosive development of the
Si cell which followed reduced costs to the point where
terrestrial applications began to appear in the early 1970's

Turning to gallium arsenide, a photovoltaic effect
was first described in it by Welker in 1954[49], followed
one year later by Gremmelmaier[50], who obtained ~1%
efficiency in a polycrystalline *p–n* homojunction cell. The
first efficient (>6%) GaAs cell was that grown by Jenny *et
al.*[51]

The first *p–n* heterojunction cell to be fabricated was
the lead sulfide cell[52], formed by the contact of two
pieces of lead sulfide, one enriched in lead and the other
in sulfur. Because of the narrow bandgap of PbS (~0.3 eV),
this cell is of interest only as an infrared detector.

The first heterojunction solar cells useful for solar-
energy applications were based on cadmium sulfide. A
photovoltaic effect was observed in CdS by Reynolds *et
al.*[53] while testing CdS rectifiers. This initiated an
intense interest in CdS-based heterojunction cells, most of
which are CdS/Cu_2S cells. The most recent type of solar
cell to be developed is the amorphous silicon cell which
dates from the mid 1970's. In 1975, Spear & LeComber[54]
became the first to produce a stable amorphous Si material.
A year later, Carlson & Wronski[55] reported a cell with a

2.4% conversion efficiency followed by Carlson *et al.*[56] with a cell with a 5.5% conversion efficiency. Following this initial work, very rapid development of this cell technology has taken place, principally in Japan.

5 EARLY SEMICONDUCTOR PHOTOELECTROCHEMISTRY

These developments in the theory and fabrication of solid-state photovoltaic cells enabled parallel advances in the understanding of photoelectrochemical effects at semiconductor electrodes. Veselovskii[57] carried out extensive photoelectrochemical investigations of Ag|AgBr, oxidized Zn, Fe, Pb, Ag, Au and Pt. By measurements of spectral sensitivity and quantum yield, he showed that the effects were produced by a process occurring in a fairly thick oxide layer, and were not surface reactions. He explained the oxidation of water to hydrogen peroxide on irradiated colloidal ZnO from experiments on oxidized Zn: absorbed radiation excited electrons to the conduction band which moved to the interior of the metal while the holes oxidized water.

Photoelectrochemical effects were observed with various other semiconductor electrodes such as cuprous oxide[58], oxidized lead[59] and selenium[60], and chlorophyll-sensitized ZnO, CdO, PbS[61]. However, the current-voltage relationships characteristic of semi-conductor electrodes were not well understood until the publication in 1955 of Brattain and Garrett's seminal work[62] on *n*-type germanium electrodes, carried out at Bell Laboratories contemporaneously with Bell's development of the silicon solar cell. Brattain and Garrett realized that the resistance of a semiconductor electrode would be much higher than that of the solution and therefore that applied bias would cause the potential difference across the space-charge layer, rather than that across the solution double layer, to change. They therefore supposed that by altering the electrical bias across the germanium, it would behave as an ideally polarisable interface and that they could charge its space-charge layer up as though it were a condenser. To their annoyance, they found that substantial currents flowed when they changed the bias, and that these were accompanied by chemical changes in the electrode surface. However, from these observations they extracted and explained the characteristic rectifying behaviour of semiconductor electrodes. Thus began a field which has combined solid state physics, electrochemistry and surface science.

Following Brattain and Garrett's work, many of the concepts of modern semiconductor electrochemistry were introduced: the relation between the sign of the photopotential and the conductivity type of the electrode,[63] the flat-band potential,[64] semiconductor electrode surface states,[65] photocorrosion.[63,66] The suppression of semiconductor corrosion by the addition of an appropriate redox species to the solution, which has proved particularly important in the case of cadmium selenide and its stabilisation by the presence in solution of the polysulphide ion, was proposed in an Electrochemical Society discussion by Geoffrey Barker[67] in 1966.

Perhaps the most important potential application of semiconductor photoelectrochemistry is to the photoelectro-lytic splitting of water to produce oxygen and hydrogen. Unusually, this made its debut in fiction, in Jules Verne's novel L'île Mysterieuse: *"... je crois que l'eau sera un jour employee comme combustible, que l'hydrogene et l'oxygene, qui la constituent, utilisee isolement ou simultanement, fourniront une source de chaleur et de lumiere inepuisable. Je crois donc que lorsque les gisements de houille seront epuises, on chauffera et on se chauffera avec de l'eau. L'eau est le charbon de l'avenir."*

Returning to the factual domain, water splitting was mentioned by Hillson & Rideal[68] as early as 1949. In 1968, Boddy[68] observed photocurrents due to oxygen evolution on n-TiO$_2$ under bias, and in 1969, Beckmann & Memming[70] reported hydrogen evolution on illuminated biased p-Gap.

In the early seventies, Fujishima and Honda[71] made the key point that titanium dioxide eletrodes were essentially stable over long periods of use as photo-anodes, and ushered in the modern era of semiconductor photoelectrochemistry by suggesting that such systems could be used for sustained solar energy conversion. There was an induction period of some two years after this first English language publication of their work, but by 1975 a flood of papers was hitting the journals. That is where my story ends.

6 POSTSCRIPT

In 1951, a UK Expert Committee opined that the efficiency of photovoltaic cells would never exceed 1%. Then came the Bell Laboratories announcement of the 6% efficienct cell, and today silicon cells of efficiency greater than 20% are

available. In the early seventies, the new community of
photoelectrochemists was confident that their devices of 1%
efficiency or so would be capable of similar improvement in
performance. To a certain extent, their optimism has been
justified in that photoelectrochemical devices of some 15%
efficiency have been made. However, it has not proved a
simple matter to do so. The complexity of some photoelectrochemical systems, their vulnerability to corrosion
and their resulting short operational lifetimes have proved
formidable barriers to commercial development. To a
certain extent, the technology now looks for uses beyond
the small-scale and specialised applications - to
photoetching and high-temperature photothermolytic waste
destruction - already under development. It seems that the
jury will stay out on photoelectrochemical solar energy
conversion until the oil wells finally run dry.

References

1. V.A. Myamlin and Yu.V. Pleskov, 'Electrochemistry of
   Semiconductors', Plenum Press, New York, 1967.
2. Yu.V. Pleskov and Yu. Ya. Gurevich, 'Semiconductor
   Photoelectrochemistry', Consultants Bureau, New York,
   1986.
3. 'Semiconductor Electrodes', ed. H.O. Finklea,
   Elsevier, Amsterdam, 1988.
4. E. Becquerel, Compte Rendu, 1839, 9, 145-149.
5. Bearing in mind the high ohmic resistance of
   Becquerel's arrangement, the distinction between the
   reacting species and the current-carrying species, and
   the logarithmic nature of the Nernst equation, we
   should be wary of this assertion, though there are
   circumstances (e.g. when the current is mass-transport
   controlled) in which it is correct.
6. J.B. Biot, Compte Rendu, 1839, 9, 169-173.
7. E. Becquerel, Compte Rendu, 1839, 9, 561-567.
8. We would say something similar, that the junction
   between the metal and the superficial semiconducting
   oxide film has rectifying properties.
9. J.B. Biot, Compte Rendu, 1839, 9, 579-581.
10. E. Becquerel, Compte Rendu, 1839, 9, 711-712.
11. J.B. Biot, Compte Rendu, 1839, 9, 719-726.
12. E. Becquerel, Ann. Chem. Phys., 1843, 9, 268; Compte
    Rendu, 1851, 32, 83; Ann. Chim. Phys., 1851, 32, 176;
    Compte Rendu, 1854, 39, 63; Ann. Chim. Phys., 1859,
    56, 99; La Lumiere II, 1868, 121.
13. A.E. Becquerel, Compte Rendu, 1841, 13, 198-200; Ann.
    Chem., 1843, 9, 257-322; 1851, 32, 176-194.

14. H. Rigollet, <u>Compt.</u> <u>Rend.</u> <u>Acad.</u> <u>Sci.</u> <u>Paris</u>, 1893, <u>116</u>,
    878-879; <u>Journal</u> <u>de</u> <u>Physique</u>, 1897, <u>6</u>, 520-525.
15. A. Stoletow, <u>Compt.</u> <u>Rend.</u> <u>Acad.</u> <u>Sci.</u> <u>Paris</u>, 1888,
    <u>106</u>, 1149-1152 and 1593-1595.
16. E.L. Nichols and E. Merritt, <u>Phys.</u> <u>Rev.</u>, 1904, <u>19</u>,
    396-421.
17. P. Hodge, <u>Phys.</u> <u>Rev.</u>, 1909, <u>28</u>, 25-44.
18. A. Goldmann, <u>Ann.</u> <u>der</u> <u>Physik.</u>, 1908, <u>27</u>, 449-536.
19. E. Baur, <u>Z.</u> <u>physik.</u> <u>Chemie</u> <u>(Leipzig)</u>, 1908, <u>63</u>, 683-
    710.
20. H. Schiller, <u>Z.</u> <u>physik.</u> <u>Chem.</u> <u>(Leipzig)</u>, 1912, <u>80</u>,
    641-669.
21. G. Trumpler, <u>Z.</u> <u>physik.</u> <u>Chem.</u>, 1915, <u>90</u>, 385-457.
22. T. Swensson, <u>Arkiv.</u> <u>Kemi.</u> <u>Mineral</u> <u>Geol.</u>, 1919, <u>7</u>,
    Articles 19 and 25.23. A.W. Copeland, O.D. Black and A.B
    Garrett, <u>Chem.</u> <u>Rev.</u>, 1941, <u>31</u>,177-226.24. A. Grumbach,
    <u>Compt.</u> <u>Rend.</u> <u>Acad.</u> <u>Sci.</u> <u>Paris</u>, 1923, <u>176</u>, 88-90.
25. C.C. Murdock, <u>Proc.</u> <u>Natl.</u> <u>Acad.</u> <u>Sci.</u> <u>U.S.A.</u>, 1926, <u>12</u>,
    504-507.
26. W. Rule, <u>Proc.</u> <u>Natl.</u> <u>Acad.</u> <u>Sci.</u> <u>U.S.A.</u>, 1928, <u>14</u>, 272-
    278.
27. J.C. Ghosh, <u>Z.</u> <u>physik.</u> <u>Chem.</u> <u>(Leipzig)</u>, 1929, <u>B3</u>, 419-42
28. H.W. Russell, <u>Phys.</u> <u>Rev.</u>, 1928, <u>32</u>, 667-675.
29. E.K. Rideal and E.G. Williams, <u>J.</u> <u>Chem.</u> <u>Soc.</u>, 1925,
    <u>127</u>, 258-269.
30. N. Sasaki and K. Nakamura, <u>Sexagint.</u> <u>Y.</u> <u>Osaka,</u> <u>Chem.</u>
    <u>Inst.</u> <u>Dept.</u> <u>Science</u>, 1927, Kyoto Imperial University,
    Japan, 249-254.
31. G. Holst, <u>Z.</u> <u>physik.</u> <u>Chem.</u> <u>(Leipzig)</u>, 1935, <u>A175</u>,
    99-126.
32. G. Holst, <u>Z.</u> <u>physik.</u> <u>Chem.</u> <u>(Leipzig),</u>, 1937, <u>A180</u>, 161.
33. E. Rabinowitch, <u>J.</u> <u>Chem.</u> <u>Phys.</u>, 1940, <u>8</u>, 551-559 and
    560-566.
34. J.J. Rowlette, *Research on Chemical Reactions Capable
    of Converting Solar Energy into Forms which can be
    Used as Power Sources,* 1960, ARL-TN-60-135, prepared
    under contract AF33(616)-6546.
35. R.L. Kerr, in *Performance Forecast of Selected Static
    Energy Conversion Devices,* 1967, Advisory Group for
    Aeronautical Research & Development, NATO.
36. K.M. Sancier, in *Trans. Conf. on Solar Energy: The
    Scientific Basis,* University of Arizona Press, Tucson,
    Arizona, 1958.
37. T. Kuwana, <u>Electroanal.</u> <u>Chem.</u>, 1966, <u>1</u>, 197-240.
38. W. Smith, <u>Nature</u> <u>(London)</u>, 1873, <u>7</u>, 303.
39. W.G. Adams and R.E. Day, <u>Proc.</u> <u>Roy.</u> <u>Soc.</u>, 1877, <u>A25</u>,
    113-117.
40. C.E. Fritts, <u>Am.</u> <u>J.</u> <u>Sci.</u>, 1883, <u>26</u>, 465-472.
41. W. Hallwachs,*Physik. Z.,* 1904, **5**, 489-499.
42. O. Grondahl and P.H. Geiger, *Trans. AIEE,* 1927, **46**, 357.

43. T.W. Case, *Phys. Rev.*, 1920, **15**, 289-292.
44. R.S. Ohl, U.S. Patent No. 2,402,662, 1941.
45. E.F. Kingsbury and R.S. Ohl, *Bell Sys. Tech. J.*, 1952, **31**, 802.
46. W. Shockley, *Bell. Syst. Tech. J.*, 1949, **28**, 435-489.
47. C.S. Fuller and J.A. Ditzenberger, *J. Appl. Phys.*, 1954, **25**, 1439.
48. D.M. Chapin, C.S. Fuller and G.L. Pearson, *J. Appl. Phys.*, 1954. **25**, 676-677.
49. H. Welker, H., *Physica*, 1954, **20**, 893.
50. R. Gremmelmaier, R., *Z. Naturforsch.*, 1955, **19a**, 501.
51. D.A. Jenny, J.J. Loferski and P.O. Rappaport, *Phys. Rev.*, 1956, **101**, 1208.
52. L. Sosnowski, J. Starkiewicz and O. Simpson, *Nature*, 1947, **160**, 816.
53. D.C. Reynolds, G. Leies, L.L. Antes and R.E. Marburger, *Phys. Rev.*, 1954, **96**, 533-534.
54. W.E. Spear and P.G. LeComber, *Solid State Commun.*, **17**, 1975, 1198.
55. D.E. Carlson and C.R. Wronski, *Appl. Phys. Lett.*, 1976, **28**, 671.
56. D.E. Carlson, C.R. Wronski, J.I. Pankove, P.J. Znucchi and D.L. Staebler, *RCA Review*, 1977, **38**, 211.
57. V.I. Veselovskii, J. Phys. Chem. U.S.S.R., 1941, 15, 145; 1946, 20, 269 and 1493; 1947, 21, 983; 1948, 22, 1302 and 1427; Zhur. Fiz. Khim., 1952, 23, 1095; 1952, 26, 509.
58. V.E. Kasgkarev and V.M. Kosonogoya, Zhur. Eksptl. i. Teoret. Fiz., 1948, 18, 927.
59. V.I. Ginzburg and V.I. Veslovskii, Zhur. Fiz. Khim., 1952, 26, 60.
60. R.W. Pittman, J. Chem. Soc., 1953, 855 and 3888.
61. E.K. Putseiko, Doklady Akad. Nauk. U.S.S.R., 1953, 90, 1005; 91, 1071.
62. W.H. Brattain and C.G.B.Garrett, Bell Syst. Tech. J., 1955, 34, 129.
63. R. Williams, J. Chem. Phys., 1960, 32, 1505.
64. J.F. Dewald, Bell Syst. Tech. J., 1960, 39, 615.
65. P.J. Boddy and W. Brattain, J. Electrochem. Soc., 1962, 109, 1053.
66. D.R. Turner, J. Electrochem. Soc., 1961, 108, 561.
67. G.C. Barker, discussing a paper by H. Gerischer, J. Electrochem. Soc. 1966, 113, 1182.
68. P.J. Hillson and E.K. Rideal, Proc. Roy. Soc. (London), 1949, A199, 295.
69. P.J. Boddy, J. Electrochem. Soc., 1968, 115, 199.
70. K.H. Beckmann and R. Memming, J. Electrochem. Soc., 1969, 116, 368.
71. A. Fujishima and K. Honda, Nature (London), 1972, 238, 37.

Birmingham and the Beginnings of Industrial Electrometallurgy

G. F. Williams

LONDON, UK

1 INTRODUCTION

One cannot claim that the electrodeposition of metals on an industrial scale was of earth-shattering proportions or even that it compared in significance with some of the other technological innovations of the nineteenth century. Nevertheless it makes a very interesting study of the relationship between a technological innovation and the science of its time, and, of course, it was an innovation which centred on Birmingham.

The nodal event was the patent for electrogilding and silvering granted to George Richards Elkington and Henry Elkington, both of Birmingham, in September 1840. The process described involved the use of gold and silver compounds dissolved in potassium cyanide solution. Several contributing strands came together here and one can begin to tell the story starting with any one of these strands.

On the demand side there was an opening for the new process because of the shortcomings of the existing gilding and silvering processes. On the production side, the belief that one could consistently obtain firm, coherent layers of metal by electrodeposition only developed in the late 1830s, and even when suitable electrogilding and silvering techniques were discovered around 1840, development would have been limited without improved techniques of nickel extraction, to provide sufficient German silver as the white base metal to be plated. Of crucial importance too were the entrepreneurial skills of the Elkingtons and Josiah Mason, the risk-takers who provided the capital and the technical and organisational expertise needed to see the process through into successful industrial production.

2 EARLIER METHODS OF GILDING AND SILVERING

Metals were gilded previously using gold amalgamated with mercury. The amalgam was spread on the surface to be gilded and the mercury driven off by heat. The hazards of mercury poisoning were considerable, and not only to the gilder. An account appearing in the Philosophical Magazine in 1801 described mercury pollution in Birmingham as follows:

> "Thus a principal part of the mercury ascends the chimneys, is deposited on
> the tops of houses and about the adjacent neighbourhood, and great
> quantities are inhaled and absorbed by the operator, keeping him in a state of
> salivation till the disease obliges him to desist.
> "Considerable quantities of mercury thus volatilised are found united and
> collected in small pools in the spouts and gutters on the tops of buildings.
> Thus many tons of mercury have been dissipated about the town and
> neighbourhood of Birmingham, to the great injury of the inhabitants. The
> poor sweep who has ascended the chimneys has been salivated, and the
> manufacturer has sustained considerable loss !" [1]

Well before the introduction of electrogilding, equipment was devised to reduce
mercury loss and poisoning. An arrangement which was certainly used in 1801 by M.
Sanders, an eminent Birmingham button maker,[2] is shown in Fig. 1. * It was described as
partially successful. The bulk of the mercury was recovered and the health of the operator
was greatly protected.

By 1829 the situation in Birmingham was sufficiently improved for Dr. Darwal to write :

> "The only other disease which it appears necessary to mention, as depending
> on the trades in this place, is what is called among the common people, the
> Shakes, or the Gilders' Palsy. The improved modes of button gilding have
> made this a much rarer complaint than formerly, and even in the toy gilding,
> which is still executed with the cap and pan, the improved construction has
> diminished the evil." [3]

Nevertheless, despite what Dr. Darwal had to say, in many cases the available apparatus
was not applied because of the expense and because it slowed down the work of the gilder.
Health hazards were not eliminated and they certainly motivated the Elkingtons in their
search for new methods of gilding using aqueous solutions.

The most important process for silver plating existing at this time was Old Sheffield
plating. This involved fusing a layer of silver onto an ingot of copper and rolling out the
ingot to obtain a sheet of plated copper. Articles were then made by shaping the plated
metal. This obviously restricted the designs achievable. The human form, for example, was
never attempted. Also there were unsightly edges which had to be covered up and which
tended to show with wear.

When commercial electro-silver plating developed in 1840, German silver, already
increasingly used in plated ware , was quickly adopted as a white metal base. Whereas Old
Sheffield plate based on German silver was much more difficult to shape than that based on
copper, for electroplate the German silver could be cast or wrought into the required shape
before the silver coating was applied.

*Figures appear on pp. 172-179

3 DEVELOPMENTS IN ELECTRODEPOSITION

Volta's pile, Fig. 2, was invented in 1800. [4] It was the first source of a continuous electric current. The deposition of metals by electrolysis was discovered very soon after and an electrogilding process due to Professor Brugnatelli of Pavia University in Italy was reported as early as 1804 in van Mons's Journal and in 1805 in the Philosophical Magazine. He had gilded two silver medals

> "by bringing them into communication by means of a steel wire with the negative pole of a Volta pile, and by keeping them, one after the other immersed in ammoniuret of gold, newly made and well saturated." [5]

However, nothing came of this and experts in the 1840s, investigating the technique described by Brugnatelli, stated that industrial gilding by this method would have been impracticable. [6]

Although a few accounts of the deposition of firm metallic layers can be found, the general experience up to the late 1830s was that, in electrolysis, metals were deposited as crystalline growths with very little adhesion to the electrode.

Before 1838, the nearest approaches to commercial electrodeposition were by Henry Bessemer, later to invent the steel converter, and by A. C. Becquerel in France. Early in the 1830s Bessemer had coated metal objects with copper by standing them on zinc trays in copper sulphate solution and had developed a technique for giving the deposit a green-bronze appearance. However, he chose to pursue other enterprises which he thought more profitable at the time, and, in his own words, "lost a grand opportunity." [7]

In 1835-6, Becquerel constructed a plant for the electrolytic extraction of silver, lead and copper at Grenelle in France. He operated this for many years but never on an economic basis. [8]

As it turned out, the line from science to economically successful electrodeposition of metals lay through battery development and the invention of the Daniell cell in 1836.

Up to about 1820, scientists seem to have been concerned to build batteries of ever-increasing size, with larger numbers of plates or plates of greater area, and to find out what these could do. For example, Pepys's battery (1823) consisted of zinc and copper plates, each about 60 foot by 2 foot, coiled round a cylinder of wood and separated from each by pieces of rope. The whole coil was immersed in a tub containing about 55 gallons of acid (Fig. 3). [9]

After Oersted's discovery, in 1820, of the magnetic effect of an electric current and the construction of galvanometers, experimenters turned their attention to the conducting wire, to the effect of using wires of different metals, of different cross-sectional areas and of different lengths. This could be only be done successfully using a constant current. There was thus a demand for a constant, non-polarising battery which Daniell satisfied with his battery invented in 1836 (Fig. 4). [10]

The conditions in Daniell's battery were such that copper was deposited as a firm metallic layer, and it was quickly noticed by a number of people, including Daniell himself, that if the layer of deposited copper was peeled off, it had the marks of any scratches on the original electrode reproduced on it in relief. By 1838 the possibility of using this for producing printing surfaces and for electroforming other objects was perceived. Priority for this goes to M. H. Jacobi in Russia. [11] There were rival claims from C. J. Jordan, [12] a London printer, Thomas Spencer, [13] a carver and gilder of Liverpool, and J. Dancer, [14] a Liverpool scientific instrument maker. Spencer's claim seems to have no justification but he made a considerable contribution to the development of electroforming, [15,16] and one can see that the conditions of knowledge were such that a number of experimenters were feeling their way independently in the same direction at the same time. One of Spencer's important contributions was on techniques for making non-conducting surfaces conducting. Thus moulds taken of works of art could be made conducting and then copies obtained by electroforming.

The early apparatus for electroforming operations consisted of rearrangements of the Daniell cell (Fig. 5).

The introduction of electroforming in 1839-40 caught the public imagination and it provided a hobby for amateurs as well as taking the attention of those who might be professionally interested. Prince Albert, who was interested in things scientific, was reported to have had his own electroforming kit. One result of the discovery was to give people faith that obtaining firm coherent deposits was possible and to stimulate the search for effective methods of electrogilding and silvering. For example, Auguste de la Rive in Geneva, who had tried electrogilding in 1825 without much success, took up his investigations again in 1839-40. He described, in 1840, a method using a solution of gold chloride, as neutral as possible. But this process was unacceptable to gilders because of poor adhesion and because it did not give the deep yellow colour required. [17]

It is within the context of this widespread interest in electroforming and the hope it gave for successful electroplating, that one must consider the events in Birmingham around 1840.

At about the same time as de la Rive was developing his process, John Wright, a Birmingham surgeon, was experimenting independently with solutions of silver and gold, trying to obtain thick, coherent deposits of these metals, comparable with the copper deposits then available. Success eluded him until the day when he finished his last, unsuccessful experiment and left his small laboratory. He had decided not to do any further work on the subject, sat down in his library and became interested in Scheele's 'Chemical Essays', written over fifty years before. There, in Essay 22, 'Experiments on the colouring matter in Berlin or Prussian Blue', he found a passage which provided the solution to his problem. [18] It referred to the solubility of gold and silver compounds in alkali cyanide solutions.

Wright immediately prepared a solution of silver chloride in potassium ferrocyanide solution, having no simple cyanide available, and using this solution as electrolyte, he

obtained thick, adherent deposits of white silver metal. Within a few weeks he had
perfected solutions of silver and gold, using potassium cyanide solution, which laid the
foundation for commercial electroplating.

The first article that received the successful coating was a small vase and the next was a
small figure of a kid. He used a common, porous garden pot containing the silver solution,
this being placed in dilute sulphuric acid in an outer vessel. The article to be coated was
placed in the silver solution and connected by a wire with a cylinder of zinc surrounding the
porous pot and immersed in the acid. [19]

According to T. H. Rollason, a nephew of Mrs. Wright, this discovery was made in 1839.
In a letter written many years later, Rollason stated :

> "In 1839, as a schoolboy at King Edward's College, I was visiting Dr. Wright's
> home, 122 High St., Bordesley, and perfectly recollect when one morning at
> breakfast he showed my aunt, Mrs. Wright, a metal plate he had just silvered
> and a brass metal chain he had gilt by the electro-process he had just
> invented. He was in high glee by his success." [20]

At the end of August 1840, Wright went to London to consult a patent agent on the
drawing up of a patent specification for his process, and while there he became acquainted
with George Richards Elkington who was in London for a similar purpose.

George Elkington, and his cousin, Henry Elkington, had applied for a patent on
improved methods plating metals on 25th March 1840. Their specification seems to have
included electro-silver plating using ammoniacal solutions of silver oxide.

The result of this meeting, and subsequent negotiations, was that Wright's process was
incorporated into a patent granted to G. R. and H. Elkington on 25th September 1840, and
entitled "Improvements in Coating, Covering or Plating certain Metals". [21] This enabled the
Elkingtons to exercise very considerable control over the electroplating industry during the
next fourteen years.

4 THE ELKINGTONS [22]

George Richards Elkington (1801 - 65) was already in the gilt toy trade. He was the son of
James Elkington, a gilt toy and spectacle maker of Bishopsgate, Birmingham. In 1815 he
was apprenticed to his mother's brothers, Josiah and George Richards, gilt toy makers of St.
Paul's Square. He showed considerable business ability and eventually, in December 1824,
he was taken into partnership by George Richards, who had establishments at 43 St. Paul's
Square and in London. In 1829 Elkington took over the Birmingham business entirely while
his uncle retained the London business.

George Elkington was also experimenting with gilding using aqueous solutions. On 24th
June 1836 he was granted a patent for gilding by simple immersion in alkaline gold
solutions. [23] A solution of gold in aqua regia was treated with a quantity of potassium

bicarbonate in large excess over that required to neutralise the solution and then take the gold up into alkaline solution. The articles to be gilded were perfectly cleaned, suspended on wires and dipped by workmen for up to a minute in the boiling gold solution.

Henry Elkington (1810 - 52) was George Elkington's cousin and also his brother-in-law. On leaving school he was apprenticed in the gilt toy trade, to George Elkington's father. In the 1830s he was in business on his own account in Camden St. as a manufacturer of moulded metal goods. The immersion gilding patent taken out by George Elkington in England was patented by Henry Elkington in France on 15th December 1836. [24] In the preamble to this French patent Henry Elkington specifically stated that a desire to avoid the hazards of the amalgam process was a motive for the invention.

The Elkingtons also took out patents for silvering by simple immersion in aqueous solutions, [25] but these processes were never of any commercial significance. They patented several other processes for coating metals with other metals. Notable among these is one patented jointly by O. W. Barratt and George Elkington in July 1838 for coating copper and brass with zinc. [26] It was not described as electrodeposition, but to get the more electropositive metal, zinc, to deposit on the copper or brass, the object to be coated was attached to a strip of zinc and immersed in a neutral zinc chloride solution.

In February 1837, George Elkington entered into partnership with four Birmingham button-making firms to work the patent on gilding by simple immersion. [27] A new works and showroom was built for this partnership in Newhall St. in 1839. The Birmingham Museum of Science and Industry is now on the site of this building, part of which still stands.

This partnership was able to acquire the services of some experienced and skilful gilding workers, who came over from the old amalgam process. Notable among these were the Millwards, a then long established family of gilders. Other experienced employees were O. W. Barratt, Thomas Fearn and, most important, Alexander Parkes. [28]

The exact date on which Parkes entered George Elkington's employment is not known, but he was certainly there in 1839-40. The son of a brass lock manufacturer, Parkes was apprenticed as an art metal worker to a firm of brass founders, Messenger & Co., where he learnt the art of casting in plaster, wax, sulphur and metals. During his apprenticeship he became an expert in drawing, painting, wood-carving and modelling. Much of his time was spent in statuary modelling and casting life-size plaster busts.

After serving his articles with Messenger & Co., Parkes became head of the casting department at Elkington's. He was a prolific inventor with 66 patents in 46 years, among them the invention of celluloid.

By the late 1830s, then, George Elkington, an able and experienced businessman in the gilding trade, had in his employ some very able, skilful people. It is known that in 1838-40 they, themselves, were investigating the possibility of electro-silver plating. [29]

5 THE INTRODUCTION OF ELECTROGILDING AND SILVERING

Having obtained their master patent of 1840, incorporating John Wright's use of cyanide solutions for electrogilding and silvering, the Elkingtons had to turn the process to commercial advantage. The largest market for electroplate would be in London and their first move was to make an agreement with Benjamin Smith, a London silversmith who had shown an interest in their efforts to devise an electrosilvering process in 1839-40. Smith was a potentially valuable asset both for his ability to provide designs and for his knowledge of the London trade in gold, silver and plated ware.

There was the possibility of licensing the new process to Old Sheffield plate manufacturers. In 1841 Elkingtons approached Roberts, Smith & Co. and J. T. & N. Creswick, both of Sheffield. Both firms were willing to adopt the new process but not on the terms offered at that time by the Elkingtons. The Sheffield platers saw it merely as an auxiliary to their existing processes.

In the event the Elkingtons decided to greatly expand their own manufacturing capacity and, to facilitate this, extra capital was obtained by taking Josiah Mason, whom George Elkington had known since 1839, into partnership. The partnership was formed between George Elkington, Henry Elkington and Josiah Mason on 29th March 1842 with a total capital of £15000 (£6000 from George Elkington, £4000 from Henry Elkington and £5000 from Josiah Mason.)

Josiah Mason (1795 - 1881) [30] was a very able businessman and a very important addition to the firm. He had already made a fortune as a manufacturer of steel pens. He made a number of fortunes during his life, one of which was in electroplating. He built a science college in Birmingham in 1880, Mason College, and in 1900 this became part of the new University of Birmingham. He was knighted in 1872.

In 1842 Mason embarked on electroplating with the Elkingtons despite warnings received from others. In a note to his biographer he wrote :

> "My connection with Mr. Elkington alarmed my dear and best friends, as they thought certain ruin would result from such untried speculations. Many platers on the old system called upon me. I certainly had no idea that I would receive so much good advice from people I scarcely knew, even by name." [31]

In the new enterprise, the day-to-day running of the business, including correspondence and public relations, was in the hands of George Elkington. Together with his cousin, Henry, he was also responsible for the production of works of art which were important in securing the reputation of the firm. Mason helped with the business and manufacturing organisation.

The Newhall St. works was acquired by the new partnership and it is claimed that Mason laid out the general scheme of the works and was particularly interested in improving mechanical processes. The electroplating rooms were relatively small, most of the space being given over to the manufacture of articles for plating and to finishing operations. The number of employees in 1843 was three hundred.

A works in Brearly St. was used for the manufacture of spoons and forks. They were made from German silver and transported to Newhall St. for plating. In 1851, the efficiency of the Brearly St. works was greatly increased by the introduction of machinery devised by Alfred Krupp. According to Josiah Mason's biographer, it was with the money obtained from this deal that Krupp started his works in Essen, and more than once he offered Mason a partnership which Mason refused. [32]

6 THE INTRODUCTION OF MAGNETO-ELECTRIC MACHINES

An important innovation was the introduction of magneto-electric machines for generating electricity for electroplating and electroforming operations.

Faraday discovered electromagnetic induction in 1831 and the first magneto-electric generator was constructed in France, in 1832, by Pixii. Others were devised by Saxton in 1833 and by E. Clarke, a London instrument maker, in 1834 (Figs. 6,7,8.) These were all small, hand-driven machines. In Pixii's machine, the magnet rotated. In the others the coil rotated. In 1833 a Pixii machine was fitted with a commutator designed by Ampere, thus enabling a direct current to be obtained.

In the 1830s several experimenters brought about electrolysis using magneto-electric machines. However the credit for the application of magneto-electric currents to large scale electrodeposition must go to John Woolrich and his son John Stephen Woolrich.

John Woolrich (Senior) was a lecturer in the medical school in Birmingham from 1828. [33] He gave a general course on chemistry and pharmacy and, in the light of his later activities in electroplating, it is interesting to note that his syllabus included much electrochemistry, the chemistry of silver and gold, and of sulphurous acid and sulphites. [34]

There is evidence that Woolrich was experimenting on the electrodeposition of gold from ammoniacal solutions in 1834, and, helped by his son, on the electrodeposition of metals using magneto-electric machines in 1836. In 1839 they experimented on the electrodeposition of silver and this eventually bore fruit in one of the key patents of electro-silver plating, taken out on 1st August 1842 in John Stephen Woolrich's name but in fact partly owned by his father. [35] This claimed the use of magneto-electric machines for electroplating and the use of sulphite solutions for electrosilvering.

The magneto-electric machine described in Woolrich's patent is shown in Fig. 9. In principle it is a large-scale version of Saxton's machine, driven by a steam engine. The first magneto-electric machine to go into commercial operation was at the electroplating works of Thomas Prime in Birmingham in 1844. This was constructed by Woolrich to a different pattern (Fig. 10). [36]

Elkingtons gained control of Woolrich's patent in 1845. They saw large magneto-electric machines facilitating the economic deposition of copper and zinc. They employed Woolrich to build a large machine for them and this was in operation in the Newhall St. works in 1847. It had eight horseshoe magnets (Fig. 11). [37]

7 THE SPREAD OF ELECTROPLATING AND THE DECLINE OF THE OLD METHODS [38]

The firm, 'Elkington and Mason', recorded a loss only in 1842, its first year of operation. In the early years the partners took very little out of the business, ploughing back profits to finance expansion. In addition, they licensed their processes to other manufacturers. By 1853 forty-five such licences had been granted.

Amalgam gilding gave way to the electro-process very soon. Old Sheffield plating firms began to give way after about 1846 and by the middle 1850s Sheffield plating had come virtually to an end.

REFERENCES

1.   Collard and Fraser, Phil. Mag., 1801, 9, 18.

2.   Collard and Fraser, Ibid, 19.

3.   J. Darwal, Midland Medical and Surgical Reporter, 1828-9, 1, 152.

4.   A. Volta, Phil. Trans., 1800, 90, 403-31.

5.   L. Brugnatelli, J. de Chimie, 1804, 5, 357,
       reported in Phil. Mag., 1805, 21, 187.

6.   C. Christofle, 'Histoire de la dorure et de l'argenture électrochimiques', Paris, 1851, 205-6.

7.   A. Watt, 'Electrodeposition', London 1889, 60.

8.   A. C. Becquerel, Bibl. Univ. de Genève, 1838, 16, 143-52.
       Comptes Rendus, 1846, 22, 781-8.
       Ibid, 1854, 38, 1095 - 1101.
       A. C. Becquerel, 'Traité D'Electricité et De Magnetisme', Paris, 1855, Vol. 2, 366 - 413.

9.   W. H. Pepys, Phil. Trans., 1823, Part 2, 187.

10.   J. F. Daniell, Phil. Trans., 1836, 126, 107 - 124.

11.   The Athenaeum, 4th May 1839, 334.

12.   C. J. Jordan, Mechanics' Magazine, 1839, 31, 163-4.

13.   The Athenaeum, 26th October 1839, 811.

14.   H. Dirks, Mechanics Magazine, 1844, 40, 73-4.

15.   H. Dirks, Ibid.

16.   T. Spencer, 'Instructions for the Multiplication of Works of Art in Metal by Voltaic Electricity', Glasgow 1840.

17.   A. de la Rive, Ann. de Chimie, 1840, 73,398 - 416.

18.   C. W. Scheele, 'Chemical Essays', trans. by T. Beddoes, London 1786, 405.

19.   G. Gore, The Popular Science Review, 1863, 2, 38.

20.   'History of Elkington and Co.', II, 153. (See note 22.)

21.   E.P. No. 8447, 1840.(E.P. = English Patent)

22.   'History of Elkington and Co.', compiled by R. E. Leader in 1913 which consists of eight portfolio volumes of original documents of Elkington and Co Ltd.. These are held in the Metalwork Department, Victoria and Albert Museum. A summary of the contents of these documents is given in a typescript prepared by R. E. Leader in 1913 and held in the British Library (Catalogue No. 1881.a.48).

23.   E.P. No. 7134, 1836.

24.   F.P. No. 10717, 1836.(F.P. = French Patent)

25.   E.P. No. 7496, 1837.
      F.P. No. 9694, 1839.

26.   E.P. No. 7742, 1838.

27.   'History of Elkington and Co.', III, 12. (See note 22.)

28.   J. Goldsmith, 'Alexander Parkes, Parkesine, Xylonite and Celluloid', 1934, ff 56, fol., typescript, British Library Ref. 8233.d.6.
      Anon., 'Memoir of Alexander Parkes (1813-90), Chemist and Inventor', pp 14 (Printed for private circulation), n.d., Birmingham Central Library Ref. 501660.

29.   G. F. Williams, Ph.D. Thesis, University of London, 1976, 205-6.

30.   J. T. Bunce, 'Josiah Mason', London and Edinburgh, 1890.

31.   Ibid, 35.

32.   Ibid, 140.

33.   Anon., 'The History of the Birmingham Medical School, 1825 - 1925', Birmingham, n.d., 22.

34.   J. Woolrich, 'Syllabus of a course of lectures on chemistry and pharmacy', 1829 and 1832, Birmingham Central Library Ref. L. 48-93.

35.   E.P. No. 9431, 1842.

36.   'The Woolrich Magneto-Electric Plating Machine', printed pamphlet issued by Birmingham Museum of Science and Industry, n.d., 4.

37.   G. F. Williams, Ph.D. Thesis, University of London, 1976, 259-61.

38.   Ibid, 393 - 458.

Fig. 1.   Sander's apparatus for mercury gilding (1801).

A = still head
B = chimney
C = articles for gilding
D = fire
E = flue leading to B
F = brickwork
G = covered tub fairly
       full of water
I = second tub of water

Source: Phil.Mag. (1801) 9, 19.

Fig. 2. Volta's pile (1800).

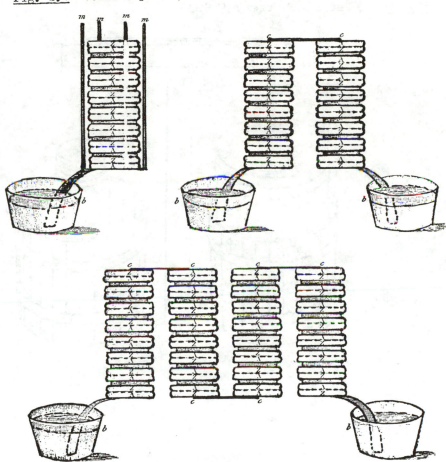

Source: Phil. Trans. (1800) 90, 430.

Fig. 3.   Pepys's battery (1823).

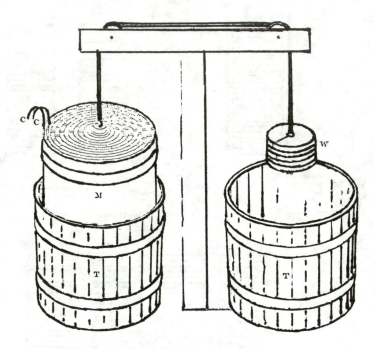

M. *Electromagnetic Apparatus.*

C C. *The Conductors.*

W. *The Counterpoise Weight.*

T T. *The Tubs, one for Acid, and one for Water.*

Source: <u>Phil.Trans.</u> (1823) Part II, 187.

Fig. **4.**   Daniell's first constant battery (1836).

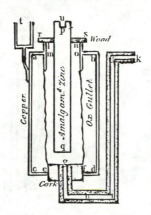

Source: <u>Phil.Trans.</u> (1836) <u>126</u>, 117.

Fig. **5.**   Spencer's improved apparatus (1840).

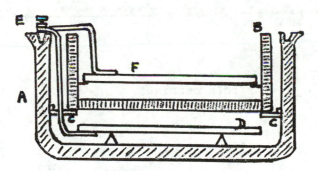

A = box of earthenware or wood
B = box (as for A) but with thick brown paper or thin unglazed
      earthenware bottom
B rests on rim cc running round the interior of A.
D = plate to be deposited on, in saturated copper sulphate
      solution.
E = binding screw.
F = zinc plate in weak zinc sulphate solution.
Source: T.Spencer, <u>Instructions on the Multiplication of Works
      of Art in Metal by Electricity</u> (Glasgow,1840), fig.20.

<u>Fig. 6.</u>   Pixii's machine.   <u>Fig. 8</u>   Clarke's machine.

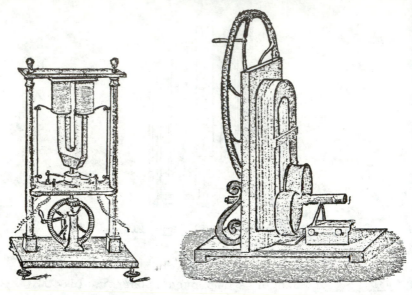

Source: R.Wormell, <u>Electricity in the Service of Man</u> (London,
        1890), p.229.

<u>Fig. 7.</u>   Saxton's machine.

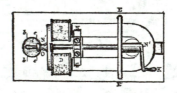

An endless cord, passing round wheel EE' and the groove of a
pulley fixed, by its centre, on the axis of N'N', communicated
motion to the coils DD. The wheel EE' was turned by the crank K.

Source: G.B.Prescott, <u>Dynamo-Electricity</u> (New York, 1884), p.532

<u>Fig. 9.</u>   Woolrich's magneto-electric machine (1842).

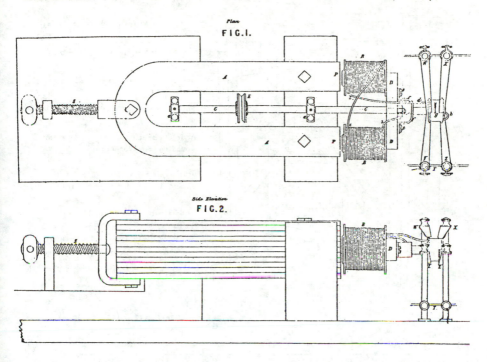

A = laminated magnet.
B = coils with solid iron cores revolving around spindle CC.
S = wooden adjusting screw.
g = commutator.

Source: E.P.No.9431 (1842)

Fig. 10. The Woolrich Magneto—electric Plating Machine. 1844.

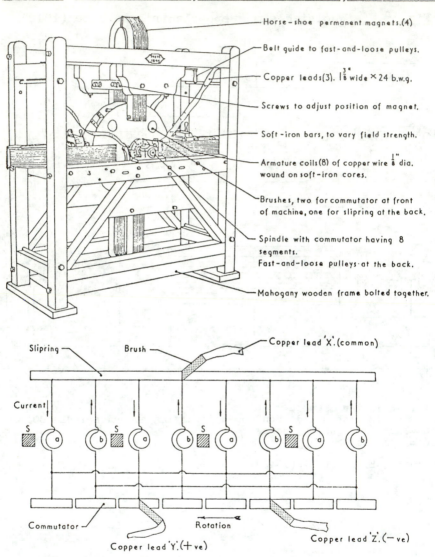

- Horse-shoe permanent magnets.(4)
- Belt guide to fast-and-loose pulleys.
- Copper leads(3). $1\frac{3}{8}''$ wide $\times$ 24 b.w.g.
- Screws to adjust position of magnet.
- Soft-iron bars, to vary field strength.
- Armature coils(8) of copper wire $\frac{1}{8}''$ dia. wound on soft-iron cores.
- Brushes, two for commutator at front of machine, one for slipring at the back.
- Spindle with commutator having 8 segments.
  Fast-and-loose pulleys at the back.
- Mahogany wooden frame bolted together.

Copper lead 'X'. (common)

Slipring

Brush

Current

S   S   S   S

a   b   a   b   a   b   a   b

Commutator

Rotation

Copper lead 'Y'. (+ve)

Copper lead 'Z'. (−ve)

Diagram of Connections.

Source: Birmingham Museum of Science and Industry.

Fig. II.   Elkingtons' machine (1847).

PLATING

Source: Illustrated Exhibitor (1851) 1, 296.

Oxygen and the Tin Can

M. Clarke

HONORARY RESEARCH FELLOW, CITY OF LONDON POLYTECHNIC, UK

1 INTRODUCTION

The preservation of food in tin cans is a technology
which is about 180 years old. The method can be used with
containers made of materials other than tinned iron or
steel sheet. Glass bottles preceded the tin can, and are
still used, but with the tin can the process achieved its
widest use. In this history, the technical knowledge of
how to make and use tools, machines, and processes, is
distinguished from the scientific knowledge of why the
technology is supposed to work as it does. Technical
knowledge, "know-how", consists of instructions and
recipes which must be followed correctly in order to
achieve the desired end. The term *operating
precepts* will be used for this type of knowledge. For
the scientific theory of why the operating precepts
succeed, the term *operating aetiology* will be used.

The logics of operating precepts and operating
aetiology are independent and different. The test of
operating precepts is - do they work? The history of
technology supports the conclusion that operating
precepts may be verified absolutely. They may become
obsolete or forgotten, but they are never falsified.
Apparent particular failure of established operating
precepts does not refute them, but is ascribed to lack of
skill, error, or accident. Scientific theories, and hence
operating aetiologies, are held to be absolutely
falsifiable, but not finally verifiable. An old
technology is likely to have had its operating precepts
explained by different, inconsistent theories at
different periods. The much-used phrase "science based
technology" is illogical, since no technology has ceased

to work when a theory explaining it has been refuted and
replaced.[1]

2 APPERT'S INVENTION

Preserving food between harvests, and for military,
exploratory, and colonial expeditions had long been
important. In 1795, all the known techniques (drying,
smoking, pickling, candying, salting, and freezing) had
major limitations. All except freezing changed the
flavour and reduced the nutrient value of foods. Freezing
was limited to areas where natural ice could be procured
and stored in ice houses. The main incentive to find new
methods was military, as has often been the case. Naval
sailing vessels were slow, and national rivalries kept
them at sea for longer and longer periods. Scurvy was a
deadly hazard to sailors who had to subsist for months on
food preserved by the current methods. Only half the crew
might survive a long voyage round the world. In 1779 the
British Channel Fleet sent 2500 men to Naval Hospitals,
and had another 1000 ill on board.[2]

In 1795 the Revolutionary Government of France
offered a prize for a better method of preserving
food,[3,4] and Nicholas Appert (1750 - 1841), a
Parisian confectioner began work on a new technique.
Thorne, in a recent work,[5] doubts whether the prize
had any influence. Appert made extensive trials, and
tested preserved foods aboard ships of the French Navy.
His new operating precepts were confirmed. The food to be
preserved, fresh and well-prepared, was cooked or
part-cooked, and placed in wide-mouthed bottles so as to
fill them. Voids were filled with juice or gravy, and the
bottles were heated in water baths to raise their
contents to near boiling. The bottles were carefully
corked to exclude all air, removed from the water and
cooled. The corks were wired in place, and for long
storage were coated with pitch or rosin. Appert proved
his method would preserve many types of food: meat, fish,
vegetables, fruit, soups, juices, and milk. Thorne tested
Apperts original operating precepts recently, and found
that they had stood the test of time.

Full details of Appert's process were published in
France by order of the Minister of the Interior, and an
award of 12 000 francs was made to the inventor. It is
extremely unusual to disperse valuable new technology
like this, especially where it has obvious military
value. Where the normative principles of science favour
full, free, and early publication of new scientific

knowledge, new operating precepts are jealously guarded
for as long as they have value. France and Britain were
at war. Napoleon had embargoed all trade between Britain
and Europe from 1806. It is astonishing that Appert's
book was immediately translated and published in London
in 1811, with a second edition (i.e. a reprint) in
1812.[6]

Operating Aetiology of Appert's Process

The important phenomenon of putrefaction had been
studied and explanations sought. During the time Appert
worked on his process, the science of putrefaction
changed as a result of the refutation of one of the major
theories of chemistry, the Phlogiston Theory.
Fermentation had three sub-classes: vinous, acetous, and
putrefactive. Macquer described 18th century fermentation
theory, noting that it was arrested by intense cold, want
of air, and want of water.[7] Macquer, like Priestley,
believed in phlogiston, and did not recognise oxygen. At
this time "air" was the generic term for the gaseous
state in Britain, though Van Helmont's term "gas" was
coming into use in Europe. Priestley called his new
discovery "dephlogisticated air".

Lavoisier refuted the phlogiston theory, and renamed
dephlogisticated air "oxygen". Many chemical phenomena
were re-interpreted. Oxygen replaced air as one of the
essential ingredients of putrefaction. To Lavoisier it
was the conversion of carbon in animal and vegetable
substances to carbonic acid, with hydrogen being evolved
as such, or as ammonia from azot-rich substances, leaving
an earthy residue. Substances containing sulphur or
phosphorus evolved sulphurated hydrogen or phosphorated
hydrogen.[8] Oxygen played a role in vinous and acetous
fermentation.[9] Gay-Lussac made critical experiments
which confirmed the oxygen theory of putrefaction. Since
in the history of chemistry Gay-Lussac is held to have
been a masterly investigator, his results commanded
respect. Robert Boyle (d. 1691) long before Appert's
birth, claimed to have preserved fruit for years inside
exhausted vessels.[10] Gay-Lussac expressed grape juice
into a bell-jar under mercury, isolated from the
atmosphere, and observed there was no putrefaction.
Oxygen prepared from manganese dioxide caused immediate
fermentation when introduced into the bell-jar.
Electrodes connected to a galvanic cell (another new
piece of technology) also induced fermentation when
introduced into the juice.[11] The new chemical theory
of putrefaction, developed contemporaneously with

Appert's process, was its first operating aetiology. The 1828 edition of Ure's Dictionary of Chemistry discussed the role of oxygen in the three classes of fermentation,[12] and explains Appert's process as depending on its exclusion.[13]

3 THE TIN CAN

Appert perfected his technique in glass bottles, being "...the most impenetrable by air."[14] He had bottles made especially for his process. Nevertheless, the disadvantages of glass were recognised in the report of the Committee which investigated Appert's claims for the Government, in 1809, and appended to Appert's book:

> With respect to the embarkation of meat necessary for a whole crew on a long voyage, a slight difficulty seems to lie in the requisite multiplicity of bottles. But Mr Appert will, without doubt, find a means to obviate this inconvenience, by the choice of vessels less fragile and of larger size.[15]

Containers made from tinned iron sheet were the solution, and were amongst the claims made in an English Patent granted to Peter Durand (Durant in some accounts) in 1810. Thorne concludes Durand plagiarised Appert's invention, taking advantage of the war, and believes Durand's patent was forgotten, though it may have been read by John Hall, one of the three founders of Britain's first cannery.[16] However, Ure states that Appert in fact commissioned Durant (sic) to seek a patent in London, and this was granted in 1811. Despite technical defects in the patent, the operating precepts were so sound that the Durant patent was purchased by the company of Donkin, Hall, and Gamble for £1000. This was worth about twice the value of Appert's French award. The company perfected the use of the tin can, and with this new container the process gradually spread world-wide.[17]

Tin and Tinplate

Food vessels made of copper or iron were tinned in antiquity.[18] Pliny records tin prevented the contamination of saline or oily food when cooked or stored in metal pots.[19] The contents tasted better, and copper did not form verdigris. Reaumur recommended the making of tinned cast iron cooking pots to promote

the French economy and to displace expensive imported
copper ones.[20] The unpleasant, sometimes fatal
effects of ingesting copper contaminated foods seems to
have been repeatedly forgotten over a period of more than
2000 years. Sweden forbade copper food vessels in its
fleet and army, and substituted tinned iron in 1754.
Others, including the Royal Navy, tinned their copper
vessels.[21] The non-toxic nature of tin corrosion
products had often been doubted because it was often
adulterated with lead, arsenic, or antimony, which
cheapened it. Tin mined in the Duchy of Cornwall had to
reach a standard quality, and the blocks carried an
official stamp. But Watson records that every tin founder
in Holland possessed counterfeit Cornish stamps, and
passed off tin as Cornish, whatever the quality.

Unlike Appert's free publication, the operating
precepts of tinplate making were guarded closely, until
the Duke of Saxony instigated industrial espionage in
about 1620.[18] The Duke passed on the secrets freely
to Yarranton in 1665, but Chamberlaine frustrated his
attempt to found a British industry.[22] Sixty years
passed before Hanbury established a tinplate industry in
Pontypool, which became pre-eminent. Tinned sheet iron
was an established commodity, known to be suitable for
food, when Donkin, Hall, and Gamble substituted the tin
can for Appert's glass bottle. The science of tinplate
was very sketchy, but this did not hinder its use.

4 THE CHEMICAL THEORY OF PUTREFACTION

Putrefaction came to be regarded as a process of slow
combustion of inherently unstable organic substances, for
which Liebig coined the term *eremacausis*.[23]
From its inception to the 1850's eremacausis was accepted
as the underlying operating aetiology.[24] A small
quantity of oxygen, unavoidably trapped in the container
when it was sealed, was removed by reaction with the food
during the high temperature stage. The subsequent
complete exclusion of oxygen removed any possible cause
of decay until the container was opened. Chemistry texts
published in the first half of the 19th century explain
Appert's process in terms of the chemical theory of
putrefaction: e.g. Turner (1834),[25] Brande
(1836),[26] Graham (1842),[27] and Wagner
(1872).[28]

Heating the contents of a can removed the flavour of
some foods. If indeed the purpose of heating were to
combine trapped residual oxygen, there were theoretical

grounds for modifying the operating precepts. Davy
suggested exhausting the oxygen with a vacuum.[29]
Bevan obtained a patent for canning in vacuo, to avoid
the high temperatures used by Appert.[30] Currie
patented an alternative process of canning in carbon
dioxide.[31] Both methods can be described as
science-based technology, but the chemical theory was
incorrect. Critical experiments which falsified it were
made by Schwann in 1836: a decoction of meat screened
from the atmosphere but supplied with calcined air which
could reach it only via a heated tube, did not putrefy.
Schwann inferred putrefaction was caused by something in
air which was destroyed by heat, and not by oxygen. The
experiment was repeated by Ure, Helmholtz, and Pasteur.
Schwann put forward the theory that the atmosphere
contained minute animals and plants capable of developing
on a nutrient, but which were killed by
heat.[32,33]

Schwann's experiment failed to convince the
proponents of the oxygen theory. Liebig defended it
stoutly as late as 1851.[34] He re-stated the evidence
for eremacausis, and the conviction that its prevention
was the purpose of Appert's process. In 1843 Liebig
attributed the invention to Appert, and the science to
Gay-Lussac, but by 1851 he attributed both technique and
science to Gay-Lussac. The process was:

> ...one of the most important contributions to the
> practical benefit of mankind ever made by science,
> and for this we are indebted to Gay-Lussac.

The microbiological theory of fermentation was scorned by
Liebig as a superstition which was taking science back to
its infancy. He refutes the germ theory by its logical
inconsistency, since if putrefaction were caused by
animicules, they would, being living things themselves,
be subject to their own effects, and would destroy
themselves. The fact they were occasionally observed was
pure coincidence. Why, scoffed Liebig, the whole
atmosphere must be filled with germs; clearly a
ridiculous concept.

Failure of the New Technology

Forty years after Appert published his operating
precepts, a supplier to the Royal Navy suffered a
well-publicised failure, having departed from them.
According to Ure:[2]

Mr Goldner, some few years ago, adopted the idea
originally conceived by Sir Humphrey Davy, of
enclosing cooked provisions in a complete
vacuum.

In 1851, a large stock of cans which had been stored for
one year was found to be largely putrid.[35] The
disaster was widely reported in the press, and a Select
Committee was appointed to conduct an inquiry.[36]
Goldner had obtained the contract by undercutting rival
firms by almost 50%, a procedure in continued approval by
public authorities for saving money, but he was not
subsequently allowed to tender.

5 MICROBIOLOGICAL THEORY

Despite Liebig's efforts, the oxygen theory lost support.
Watts, in 1864, gave a balanced review of the claims of
the chemical theory (Liebig, Berzelius, and Schmidt), and
those of the microbiological theory (Schwann, Blondeau,
and Pasteur).[37] Watts concluded that the matter was
not finally settled, but the balance was towards the
microbiological theory. In 1860, the fifth edition of
Ure's celebrated Dictionary, edited by Robert Hunt since
the death of its founder, mentioned that Schroeder and De
Dusch had extended Schwann's experiment, and found that
no putrefaction occurred if air were allowed to reach
cooked food only via a fine filter.[38] Under the new
theory, the operating aetiology of Appert's process was
that the heating stage sterilised the contents of the
container, killing the fungi and germs which were
everywhere present. If, then, the container were
hermetically sealed, there was nothing to cause
putrefaction so long as the seal was maintained. This is
essentially still the operating aetiology of the process.
It is interesting to recall that Appert himself did not
subscribe entirely to the oxygen theory, and warned
against the idea that a vacuum would suffice:

It might on first view of the subject be thought that
a substance, either raw or previously acted upon by
fire, and afterwards put into bottles, might, if a
vacuum were made in those bottles and they were
completely corked, be preserved equally well (as)
with the the application of heat in the water bath.
This would be an error, for all the trials I have
made have convinced me that the absolute privation of
the contact of external air (the internal air being
rendered of no effect by the action of heat), and the
application of heat by means of the water bath, are

both indispensable to the complete preservation of
alimentary substances.

Elsewhere, Appert suggests that the application of heat
destroys the predominating agency of fermentation.

Growth of the Canning Market

For its first fifty years, the tin can was largely part
of military technology. The Royal Navy canned about 500
tons of meat each year. Canned food played a large part
in the Crimean War (1854-1856),[39] and the American
Civil War (1861-1865).[40] In the Great Exibition of
1851, Gamble exhibited a can of mutton prepared in 1824
for an Arctic Expedition, recovered in 1833, and still in
good condition.[30] Up to 1860, canned foods did not
penetrate the civil market to any great extent, because
in normal times unpreserved foods were cheaper. After
1860 however, canning enabled North America, Australia,
and New Zealand to exploit their much lower farming
costs, and to supply the European civil market at
competitive prices.[39]

6 SCIENCE AND TINPLATE

A barrier layer theory of protection was adopted from
antiquity. A continuous layer of non-toxic tin prevented
food corroding copper or iron vessels, and being
contaminated with the products.[42] Tinning was carried
out by immersing clean, oxide-free iron sheet in molten
tin (mp 231 C). The secret of success was the use of a
flux which prevented re-oxidation on entering the hot
tin. Bishop Watson (1737-1816), Professor of Chemistry in
the University of Cambridge (1764-1771), enquired into
the quantity of tin present on tinned copper and iron. He
found 254 square inches of copper gained 0.5 ounces when
tinned, equivalent to an average thickness of 12
micrometres. English tinplate makers told him 1 lb of tin
was applied to each 28 square feet of iron sheet;
assuming both sides were coated, this is also an average
of 12 micrometres. Watson was surprised that the coating
was so thin, and felt it should be thicker to resist
corrosion by vinegar and fruit acids.[43]

It is likely that tin coatings would be scratched
during the making of cans from sheet, so that both tin
and iron could be exposed to the food inside the can. The
development of electrochemistry was contemporary with
that of Appert's process, and when Donkin, Hall and
Gamble perfected the tin can, there was considerable

knowledge of the behaviour of dissimilar metals in
electrical contact immersed in a common electrolyte.
Volta's theory that contact potentials depended only on
the metals involved had been refuted. Wollaston's view
that the cell current arose from the chemical reactions
taking place at the electrodes prevailed. The polarity of
particular bi-metal couples, or voltaic circles, was
known to reverse in different electrolytes. De la Rive
had found that tin was generally the anode to copper in
acids, but copper became the anode in ammonia
solutions.[44] The technical consequences of bi-metal
couples were recognised in some fields:

> In sheathing ships it is necessary to use bolts of
> the same metal which form the plates; for if two
> different metals be employed they both oxidate
> speedily, with the water of the ocean they form a
> simple galvanic circle.

The practical consequences had been noted before the
advent of electrochemical theory. In 1670 Sir Philip
Howard and Francis Walton were licenced by Parliament to
sheath ships with rolled lead sheet. Twenty vessels were
sheathed, but the iron bolts and rudder became wasted to
a degree never seen on any unsheathed ship.[45] Watson
suggested chemists should investigate why a metal
covering could so injure the ships. Had this been done
electrochemical cells might have originated from a
shipping problem a century earlier, rather than from the
dissection of frogs!

The importance of electrochemical coupling was
understood for zinc coated iron,[46] and Davy's
protection of copper bottomed ships by attaching anodes
of zinc, iron or tin, received attention in
texts.[47,48] However, throughout the 19th, and early
part of the 20th centuries, iron was believed to be the
anode of the iron-tin couple, since it was a fact that
the corrosion of iron, deliberately exposed by scraping
tinplate, was accelerated by saline solutions containing
dissolved oxygen. It was inferred that the tin coating on
tinplate must be, and must remain, continuous, by Percy
(1864),[49] Bloxham (1890),[50], Blucher
(1911),[51] Evans (1926),[52] and by Krohnke, Maass,
and Beck (1929).[53] Percy studied tinplate manufacture
at Pontypool, and found that 8 lb of tin was required to
coat 112 sheets measuring 14 x 20 inches.[54] This is
equivalent to an average tin thickness of 12 micrometres,
as earlier reported by Watson. Although many normal cans
must have had iron exposed inside at scratches, iron

corrosion products were not reported in foods, but 10
grains of dissolved tin per pound of food (0.14 wt %) was
typical.[55] This is not consistent with the accepted
polarity of the iron-tin couple. The discovery that tin
is the anode under the conditions found inside cans was
reported in the periodical literature in 1927,[56] and
1928.[57] However the explanations were unsound until
Hoar showed that the corrosion potential of tin, usually
higher than that of iron, was so greatly reduced in
solutions of fruit acids as to fall below it.[58]
Divalent tin formed very stable complex anions with
citric, tartaric, and oxalic acids, allowing tin to
become anodic to iron even in solutions containing
dissolved oxygen. In perchloric acid, with which tin
forms no complexes, tin was cathodic to iron. It was
concluded that the relative freedom from perforation
enjoyed by tin cans was the result of this reversal of
polarity, and that inside cans tin dissolved slowly from
a large area, protecting any iron exposed at a small one.
For 120 years, the unrecognised reversal of polarity had
averted the problem which worried Bishop Watson, once he
had recognised the tenuity of the coating. Small
discontinuities, undoubtedly present in some, if not all
cans made from tinplate, had been protected by the
sacrificial action of tin. The tin corrosion products
were unobtrusive, being tasteless, colourless, and
non-toxic.

7 ELECTROLYTIC TINPLATE

There had been no radical change in the operating
precepts for making tinplate since well before the 16th
century. There had been many detailed refinements, and a
change from iron to steel with the success of Bessemer's
process for making cheap steel. The coating thickness was
reduced, until by 1930 the average was but a tenth of
that reported by Watson and Percy, but further
improvements seemed unlikely. Hot tinning imposed a limit
to the minimum thickness, which had been reached, and the
coating was inherently irregular. Major advances in steel
strip rolling had placed on the market cheap coils about
0.9 m wide, and 1000 - 2000 m long, but continuous hot
tinning, which would best have exploited the coils, was
not mastered, and they still had to be sheared into
individual plates. To make progress, a radically new
method was tried, of electrodepositing solid tin on to
the moving strip of steel, from an aqueous solution of a
tin salt.

The electrodeposition of metals started as a

commercial technology in 1840, 35 years after Brugnatelli
had gilded silver medals using the process.[59] But tin
deposited as a sponge of loose crystals, and could not be
electroplated as a coherent coating. Smee abandoned it
after many trials. Roseleur devised what became for years
the sole effective plating bath using a pyrophosphate
system.[60] The ease of applying tin by hot dipping was
a conclusive disincentive to the development of
electroplating.[61,62] In the only commercially used
process, for tinning brass and copper pins, the current
was provided by the dissolution of the copper itself, and
no external cell was needed. The operating precepts for
this electrodeposition process pre-date the invention of
Volta's pile, and the founding of the science of
electrochemistry.[63]

The proposals to make electrolytic tinplate provided
sufficient incentive for the development of
electroplating processes capable of giving smooth,
coherent coatings at high speed.[64,65] An experimental
electrolytic tinplate plant was constructed in the USA,
in 1934, which used an acid stannous sulphate
electrolyte, inhibited by complex organic additions to
prevent the growth of spongy deposits. The electroplated
tin had a matt, non-reflective surface, and to make it
aesthetically acceptable to users familiar with
traditional shiny tinplate, the electrodeposit was
polished by nickel-silver wire brushing. Polishing was
soon superseded by flow-melting, a stage in which the
tinned strip was momentarily heated above the melting
point of tin. Surface tension produced a bright, liquid
surface, retained on cooling, while the molten tin
reacted with the solid iron to produce a layer of the
alloy $FeSn_2$, also formed in hot dipping. The
technique of flow-melting had been patented as long ago
as 1872, by Lobstein,[66] so the new process still had
old elements. The first commercial electrolytic tinplate
plant began work in the USA in 1937. The advantages of
the new process were its ability to use coiled steel
strip in continuous production, its high productivity,
and its ability to produce thinner, and more economical
tin coatings than were possible by hot dipping.[67]

The outbreak of war on the 11th December 1941,
between the USA and Japan (which captured the main
sources of tin) accelerated the adoption of electrolytic
tinplate. By 1953, the USA, which was the world's main
supplier of tinplate, was making 72% of its output by
electrolysis, although only 20% of Britain's output was
of this type. Today the output of hot dipped tinplate is

negligible.

REFERENCES

1.   M. Clarke, J.F.H.E., 1982,6(2), 3
2.   A. Ure, "A Dictionary of Arts, Manufactures, and
     Mines", Appleton, New York, 4th Ed.1857, Vol. 2, p.
     127
3.   W.H.Armytage, "A Social History of Engineering",
     Faber, London, 4th Ed.1976, p. 107
4.   A.and N.Clow, "The Chemical Revolution",
     Batchworth, London 1952, p. 569
5.   S.Thorne, "The History of Food Preservation",
     Parthenon, Kirby Lonsdale 1986, p. 31
6.   Mons. Appert, "The Art of Preserving All Kinds of
     Animal and Vegetable Substances for Several Years",
     Black, Parry, and Kingsbury, London, 2nd Ed.1812
7.   P.Macquer, "A Dictionary of Chemistry", Cadell and
     Elmsley, London, 2nd Ed.1776, entry FERMENTATION
8.   A.Lavoisier, "Elements of Chemistry", Creech,
     Edinburgh, 3rd Ed.1796, Chapter 14
9.   A.Lavoisier, op. cit., Chapters 13 and 15
10. P.Macquer, op. cit., footnote added by J.Keir
     (translator) to entry PUTREFACTION.
11. J.Liebig, "Chemistry in its Applications to
     Agriculture and Physiology", Taylor and Walton,
     London, 3rd Ed.1843, pp. 299 and 327
12. A.Ure, "A Dictionary of Chemistry", Tegg, London,
     3rd Ed.1828, p. 465
13. A.Ure, op. cit., 1828, p. 673
14. Mons. Appert, op. cit., p. 18
15. Mons. Appert, op. cit., p. 162
16. S.Thorne, op. cit., p. 37
17. A.Ure, op. cit., 1857, Vol. 2, p. 128
18. J.Beckmann, "A History of Inventions, Discoveries,
     and Origins", Bohn, London, 4th Ed.1846, Vol. 2, pp.
     206 to 230
19. Pliny, "Natural History", ca AD76, translated by
     H.Rackham, Heinemann, London 1938-1962, Book 48
20. Sr.de Reaumur, "Memoirs on Steel and Iron",
     translated from the Paris Ed of 1722 by A.G.Sisco,
     Univ. of Chicago Press, Chicago, 1956, p. 351
21. R.Watson, "Chemical Essays", Evans, London, 3rd
     Ed.1784-1786, Vol. 4, p. 149
22. S.Smiles, "Industrial Biography", Murray, London
     1863, p. 66
23. J.Liebig, op. cit., 1843, p. 291
24. J.Liebig, op. cit., 1843, p. 320
25. E.Turner, "Elements of Chemistry", Taylor, London,
     5th Ed.1834, p. 922

26. W.T.Brande, "Manual of Chemistry", Parker, London 1836, p. 1133
27. T.Graham, "Elements of Chemistry", Bailliere, London 1842, p. 716
28. R.Wagner, "A Handbook of Chemical Technology", Appleton, New York 1872, p. 564
29. A.Ure, op. cit., 1857, p. 129
30. C.Tomlinson, "Cyclopaedia of Useful Arts", Virtue, London 1852 - 1854, Vol. 1, p. 706
31. C.F.Partington, "The British Cyclopaedia", Orr and Smith, London 1835, Vol. 2, p. 416
32. J.S.Muspratt, "Chemistry: Theoretical, Practical, and Analytical", MacKenzie, London 1860, Vol. 1, p. 67
33. C.W.Vincent (Ed.), "Chemistry: Theoretical, Practical, and Analytical, as Applied to the Arts and Manufactures", MacKenzie, London, not dated, probably 1880, Vol. 1, p.606
34. J.Liebig, "Familiar Letters on Chemistry", Taylor, Walton, and Maberly, London, 3rd Ed.1851, Letter 19
35. C.Knight, "Arts and Sciences", Bradbury and Evans, London, 1866 - 1868, Vol. 1, col. 391
36. S.Thorne, op. cit., p. 60
37. H.Watts, "A Dictionary of Chemistry", Longman, Green, Longman, Roberts, and Green, London 1863 - 1868, Vol. 2, p. 623
38. R.Hunt (Ed.), "Ure's Dictionary of Arts, Manufactures, and Mines", Longman, Green, Longman, and Roberts, London, 5th Ed.1860, Vol. 1, p. 34
39. C.Knight, "Arts and Sciences", Bradbury and Evans, London, Supplement 1873, col. 1728
40. C.Falls, "The Art of War", OUP, Oxford, 1961, p. 68
41. C.Tomlinson, op. cit., Vol. 1, p. 708
42. P.Macquer, op. cit., entries TIN and TINNING
43. R.Watson, op. cit., Vol. 4, pp. 179 and 201
44. E.Turner, op. cit., p. 151
45. R.Watson, op. cit., Vol. 4, p. 194
46. T.Graham, op. cit., p. 219
47. C.F.Partington, op. cit., Vol. 1, p. 586
48. W.T.Brande, op. cit., p. 282
49. J.Percy, "Metallurgy: Iron and Steel", Murray, London 1864, p. 28
50. C.Bloxham, "Chemistry", Churchill, London, 7th Ed.1890
51. H.Blucher, "Modern Industrial Chemistry", Gresham, London 1911
52. U.R.Evans, "The Corrosion of Metals", Arnold, London, 2nd Ed.1926

53. O.Krohnke, E.Maass, and W.Beck, "Die Korrosion",
    Hirzel, Leipzig 1929
54. J.Percy, op. cit., p.725
55. A.W.Blyth, "Foods: their Composition and
    Analysis", Griffin, London, 3rd Ed.1888, p. 195
56. C.L.Mantell and W.G.King,
    Trans.Amer.Electrochem.Soc.,
    1927,51,40
57. R.H.Lueck and H.T.Blair,
    Trans.Amer.Electrochem.Soc,
    1928.54,257
58. T.P.Hoar, Trans.Faraday.Soc.,
    1934,30,472
59. A.Smee, "Elements of Electrometallurgy", Longman,
    Brown, Green, and Longman, London, 2nd Ed.1843, p.
    174
60. A.Roseleur, "Manipulations Hydroplastiques",
    Roseleur, Paris 1855, Chapter 8
61. W.G.McMillan, "A Treatise on Electrometallurgy",
    Griffin, London 1890, p. 259
62. S.Field, "The Principles of Electrodeposition",
    Longmans, London 1911, p. 210
63. R.Watson, op. cit., Vol. 4, p. 199
64. M.Schlotter, J.Electrodepositors'
    Tech.Soc., 1936,11,183
65. D.J.MacNaughtan, W.H.Tait, and S.Baier,
    J.Electrodepositors' Tech.Soc.,
    1937,12,45
66. J.Napier, "A Manual of Electro-metallurgy",
    Griffin, London, 5th Ed.1876, p. 187
67. C.Frenkel, J.Electrodepositors'
    Tech.Soc., 1946,21,129

Electrochemical Corrosion and Boiler Explosions in the 19th Century

C. A. Smith

SPA ASSOCIATES, 24 CLOISTER CROFTS, LEAMINGTON SPA, CV32 6QQ, UK

1 INTRODUCTION

The science of corrosion had its first period of rapid
advancement in the first half of the nineteenth century.
This was a result of intense and sustained scientific
interest and activity aroused by the invention of the
galvanic battery and the controversy over the nature
and source of galvanic current.

One of the most significant facts of corrosion,
namely that an originally neutral water tends to become
alkaline during the corrosion process, appears to have
been stated for the first time by Austin [1] in 1788.
Austin, however, wrongly attributed the alkalinity to
the substance which we now call Ammonia, and the same
mistake was made by Chevallier [2], who repeated Austin's
work in 1808.

2 ELECTROCHEMICAL THEORY

In 1800 H Davy [3] established the intimate connection
between the electrical effects and the chemical changes
occurring in the battery, and drew the conclusion of
the dependence of the one upon the other.

In the following years Wollaston [4] stated that it
was probable that, even in the case of decomposition
of water by uncoupled metals in acid solution,
".... the formation of hydrogen gas depends on a
transition of electricity between the fluid and metals,"
and that:-
".... iron itself has the power of precipitating copper
by means, I presume, of electricity evolved during the

solution."
Following these announcements in rapid succession,
there was a lull for some seventeen years and one of the
next comments came in an anonymous letter which was
published in Ann. Chim. Phys. in 1819 in which it was
observed that:- [5]
".... iron does not decompose water at ordinary temperature when they are both perfectly pure. But once
oxidation has commenced, by whatever cause, it can
continue by the action of water alone."
for
".... iron and its oxide should be considered as two
heterogeneous bodies of different electrical energy,
and capable, by their contact, of effecting the decomposition of water, the same as an element of copper and
of zinc in the galvanic pile." Although this discussion
was unsigned, a comment by Hall [6] shortly afterwards,
dealing with a different topic, states that Thehard
was the author of the paper.

The significance of this paper is that it suggested
that rusting was also an electrochemical phenomenon and
could well have prompted the investigations of H Davy
which led him to the same conclusion in 1826 in which
he said:- [7]
"In the rusting of iron, the oxide formed by the contact
of moisture becomes the negative surface, and exalts
the oxidation of the mass of metallic iron, and consequently extends in circles."

A further early development in the electrochemical
concept of corrosion was that of the differential
solution concentration cell proposed by Becquerel in
1827. [8] He showed that a cell may be produced by a
single metal and a single fluid of two different concentrations.

Mallet utilised this discovery in 1840 to explain
localised corrosion of an iron casting which was at the
bottom of a harbour in the mouth of a tidal river. He
stated:- [9]
"It is well known that the sea water, during the flowing
of the tide, from its greater density, forces itself
between the river water like a wedge, and slowly and
imperfectly mixes with it, hence two strata, one of
fresh or brackish water, the other of salt water below
it."

More than forty years later, Andrews thought he was
the first to discover this cause of corrosion currents,

and reported measurements of potential differences
developed in tidal streams in 1884 [10] and 1890 [11].

R Adie in 1847 [12] reported interesting experiments
on oxygen concentration cells. He placed two pieces of
zinc, or iron, cut side by side from the same sheet, in
a running current of water, the one opposed to a rapid
part of the current, the other in a still place at the
edge. When these were connected with a galvanometer,
the piece of metal in the current acted as a 'negative
plate'. He stated:- [12]
"With both plates in the still water and a tube filled
with oxygen inverted over one, the effect was the same."
Adie continued:-
"It is the greater supply of oxygen to the plate in the
current which converts it into a negative ...."
"A single plate of iron exposed to water and oxygen gas
has local differences on its surface which act in the
same way as if the iron had been in two halves and
placed in a stream in the manner described ...."

The importance of variations in concentration of
oxygen and salts in solution in promoting localised
corrosion was evident from these experiments, but the
significance of these investigations was only partially
grasped, and then forgotten. Practical benefits from
an understanding of these fundamental environmental
factors did not follow until well into the twentieth
century.

Although the electrochemical theory of corrosion
had been proposed by Wollaston in 1801 and developed
by de la Rive in 1830 it was 1903 before it was again
revived:- by Whitney [13]. Although Whitney presented
no new ideas, his paper was of considerable practical
importance. He applied the theory to the corrosion of
iron and steel at a time when the cost due to wastage
of this major structural material was assuming disturbing
proportions. For this reason, his contribution received
considerable attention, and was a stimulus to further
progress.

Whitney's view may be summarised as follows:-
According to the Nernst theory, [14] iron has a natural
tendency to pass into solution in pure water; but it
will only dissolve if there is present a second substance
in contact with the iron, having a lower 'solution
tension' than iron itself. If the hydrion concentration
exceeds a certain value, ie if the solution is sufficiently
acidic, hydrogen will come streaming from the second

(cathodic) substance as a gas; but if the hydrion concentration of the water falls below that value, corrosion
will be much slower, since the hydrogen will only slowly
diffuse away in solution. In postulating the removal
of any appreciable amount of free hydrogen in solution,
Whitney made an assumption which is now believed to be
wrong.

Hall had demonstrated the necessity of dissolved
oxygen for appreciable corrosion of iron in water at
ordinary temperature, yet this fundamental consideration
had been overlooked by Whitney. Also Whitney's theory
was inconsistent with the knowledge of polarisation
currents founded upon the work of Helmholtz. [15]

3   EARLY BOILER PROBLEMS

During the period that these scientific investigations
were being pursued, the industrial revolution was in full
progress. However, there appears to have been little
interaction between the scientists involved in these
corrosion investigations and the mechanical engineers
of that period who were thrusting the technology
forward.

Although today a variety of problems continually
arise demanding a high degree of product knowledge with
particular boilers, in the early days of the industry
even knowledge of the most general kind was very sparse.
The result was that boiler manufacturers overdesigned
in the hope that this would solve all their problems.
Their lack of knowledge of the expansive force of steam
was catered for to a certain extent in this way, but
the major problems began to arise after the boilers had
been in service for several years and the boiler material
began to waste away. Catastrophic explosions in different
types of boilers began to occur and as a consequence,
the problem had to be examined in depth. Fig 1 illustrates
a typical example.

On analysing the various investigations carried out
under a variety of conditions, (Fig 2) it is found that
the early problems with boilers associated with corrosion
can be divided into three main classes:-
Furrowing, [16] [17] [18] which is a specific form of
internal corrosion; secondly, internal corrosion which
usually takes the form of pits or general wasting of
the material and thirdly, external corrosion. These
latter cases will be considered separately.

BOILERS

Damage caused by an exploding boiler in Blackburn in 1874

Figure 1

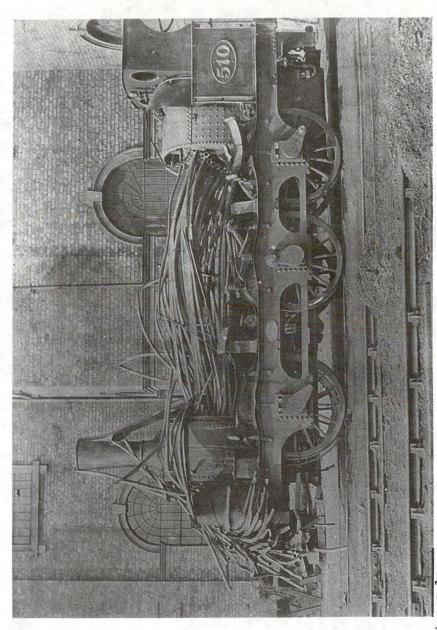

Figure 1a

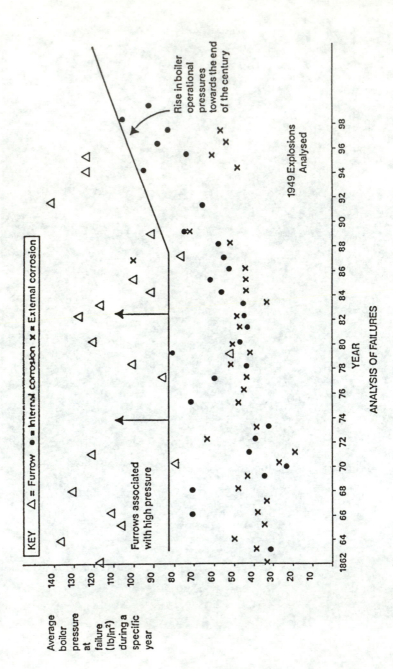

Figure 2

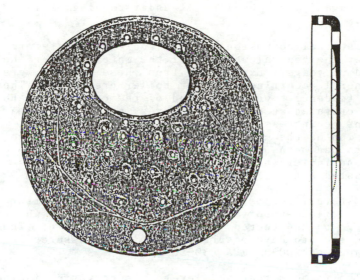

Figure 3 Illustration of general corrosion

Furrowing was associated with high cylinder
pressure and this form of corrosion very often took
place in railway engine boilers. The incidence of
explosions from this source was infrequent but accounted
for up to about 10% of cases. However, the significance
of this form of corrosion was that devastation could
be caused on a much larger scale due to the high pressures
involved and also the lack of understanding of its
cause.

Internal corrosion usually took the first form of
pits or the general wasting of the material over the
period of years. Explosions from these sources were
in general occurring at lower boiler pressures. There
was a tendency towards the end of the century for boilers
to be operated at higher pressures, and there was a
percentage increase in explosions due to internal
corrosion as the century progressed. Many of these
boilers had been operational for some twenty years or
more and failure had occurred because the boiler skin
had wasted away.

External corrosion was often due to the failure
of the boiler supports and leakage from joints which
could go on undetected for a considerable time. These
can be seen to have occurred at boiler pressures in
the range of 200-350kPa (30-50 lb/in^2).

The factors that accelerate the corrosion include
galvanic action due to lack of homogeneity or different
metals in contact; animal or vegetable grease on the
plates and indentations or scratches.

Problems of Furrowing

A report on an accident in which two men were
killed was published in 1862 in 'The Engineer'. On
examination of the wreck of the boiler, it was found
that extensive corrosion had taken place on the inside
of the middle ring of plates, and near one of the
longitudinal seams of rivets. The report stated:- [16]
".... This corrosion in grooves is known as 'furrowing',
and although well-defined cases of this sort are occasionally referred to as mysterious causes, there can be but
very few, if any, locomotive engineers who have not had
opportunities for observing the phenomenon itself. We
have ourselves seen, on several of our great lines,
(railway companies) boiler plates otherwise apparently
sound, but with a long deep groove - almost resembling
the cut of a square-nosed tool, half an inch or so from

a riveted joint."

The 'furrowing' referred to was a form of internal corrosion which was found to be a frequent cause of boiler explosion, especially in locomotive boilers.

It can well be appreciated that corrosion in this form would rapidly cause plate failure. It occurs where there is a sudden change of thickness either along the lines of the seams, or opposite the edge of angle-iron attachments. It is still a major cause of failure and is a form of stress-corrosion. The variation in pressure and temperature causes strains in the material and consequently cracks occur which when within a steam environment result in eventual failure.

An insight into the form of inspection carried out on locomotive boilers at this time by one of the railway companies was also mentioned:- [16]
".... We have understood, indeed, that on the Great Northern Line, it has lately become the practice, when an engine is in for repairs, to take out a few of the tubes next to the plates of the boiler, and after thrusting in a lamp on the end of a long rod to observe as far as maybe, the condition of the plates. But little could be gathered, however, from such an examination."

The inference from the article was that little or no further action was taken and the engine was allowed to continue, presumably until it exploded with a con- sequent loss of life.

Another article appeared in 'The Engineer' about this time, which indicated that a more thorough invest- igation of corrosive action had taken place. H W Tyler was making the examination on behalf of the Railway Department of the Board of Trade. He stated:- [17]
".... In examining the fractured portion of this boiler, I observed that corrosion had been actively going on in several parts of it, particularly above two seams of rivets and situated on the left side of the middle ring. The metal had been eaten away along a line in an irregular manner, and had been reduced in places to about $^1/16$ inch.

This value of $^1/16$ inch wall thickness mentioned by Tyler appears to have been a critical thickness for operating conditions of pressure at that time, since this value is mentioned at many subsequent inquests. It was obviously the critical thickness at which rupture might be expected at the ordinary working pressure of the engine, ie 100-120 lbs/ins^2.

Tyler goes on to say
".... The interior of the barrel had last been examined
in 1857, at which time, a new set of tubes was inserted.
It was said to have been in good order and was expected
to outlast another set of tubes ...."

It is very likely that the boiler was not examined
for furrowing at the earlier date since it is difficult
to believe that the reduction in plate thickness of
about ^{1}/6 of its original dimension had occurred in
less than six years. This fact was evident to Tyler,
who suggested:-
"It is necessary that inspection be carried out more
frequently into the conditions of plates. When no
symptoms of leakage or other indication of defect are
observed, it is at present practice in locomotive works
not to examine interior surfaces of barrels of engine
boilers except when tubes are taken out for removal.
The first set of tubes in engines last six years. All
tubes should be taken out and a boy sent into the
interior to brush away deposits formed from impurities
in the water."

This was obviously a great step in the right direction, but it is probable that the boy performing the
task would also be asked to look for any corrosion and
it is very unlikely that he would have had the skill
and experience to carry out this work.

Design Changes Introduced

During this period between 1857 and 1862, there
was evidently a change in boiler design to overcome this
problem of furrowing. [19]
"In order to remedy the tendency of corrosion along the
seams of rivets it has latterly become the practice to
place the upper plates inside the lower ones, instead
of outside them, at the joints, a mode of construction
which had not been adopted when this boiler was made
(1857). It is generally believed, that by avoiding the
ledge which is formed by the edge of the lower plate,
when that plate is placed inside the upper one, the
deposit caught by such a ledge, which promotes corrosion
is deprived of its resting place. This is only a partial
remedy since plates continue to be eaten away below the
seam in somewhat the same way."

The reference that had been made by Tyler in 1862,

to the 'ledge action' as a major factor contributing to
furrowing, was again referred to four years later by
Kirtley who illustrated his comments:- [19]
"In the longitudinal joints of the boiler, the grooving
from corrosion is generally found to be more marked when
the inside ledge of the lap faces upwards, as in E,
Fig 4), than when it is turned downwards, as at D. In
the former case, it may be considered that the deposit
will collect upon the projecting ledge in large quantities, forming a thickness of deposit sufficient to be
detached bodily by the springing of the plate under
pressure; and it will consequently leave the bare plate
more frequently and extensively exposed to the direct
action of water, than when the edge of the plate faces
downwards, as at D because in the latter case, the thin
film deposit will not be so readily and frequently
detached from the plate by the same action."

    In the four years that passed since Tyler's paper,
the problems of furrowing received more extensive investigation and Kirtley was able to outline the problem in
a more precise way than any of the earlier investigators;-
[19]
".... In the present ordinary construction of locomotive
boilers with lap joints, the wear by corrosion of the
plates is found principally round the smokebox end of
the boiler barrel, in the interior, opposite to the edges
of the outside angle iron, where an angular groove is
found to be eaten out of the plate by corrosion."
"In the ordinary construction of locomotive boilers with
lap joints, the boiler is constructed of three rings,
each ring formed by two plates of $^7/16$ inch thick,
riveted with lap joints. The general amount of lap is
$2\frac{1}{2}$ inches for single riveted and $3\frac{1}{2}$ inches for double
riveted joints. The smokebox and firebox are each
united to the barrel of the boiler by an angle iron KK.
General experience has shown that after five or six
years wear of these boilers, the grooving action is
developed at the joints and at the edge of the angle
iron rings."

    Now the longitudinal strain upon the joints of
boilers constructed in this manner tends to spring and
bend the plates at the joints, when under pressure,
Kirtley then goes on to explain that the strain acts
in the direction SS springing and bending the plates.
He continued:-
".... The continued alternation of expansion and contract-

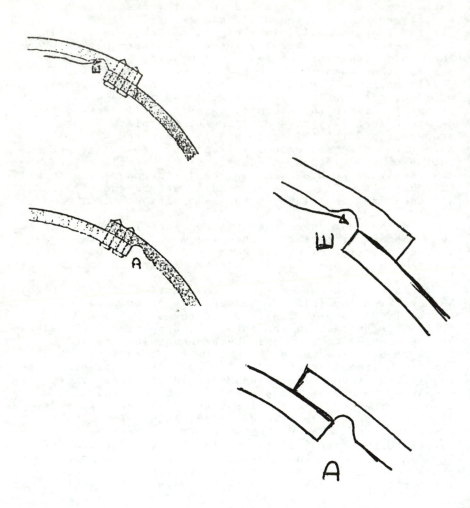

Figure 4

tion in the boiler causes the scale that is deposited
upon the plates from the water to be continually broken
off at the edge of the joints by the mechanical action
of this springing and bending of the plates at the lines
of the joints; and the plates are thereby laid bare at
these parts and kept continually exposed to the corroding
action of the water, instead of the water by the deposited
scale remaining attached to them. Though the corrosion
produced by the water is slow in action and slight in
effect on the rest of the boiler plates, which are
protected by some deposit of incrustation remaining
almost constantly upon them, it becomes very serious on
an exposed surface of iron ...."

"It must further be noticed that the pressure, under
which the locomotive boilers are worked, is much higher
than is the case of stationary boilers and the injurious
action caused by the springing of the plates at the
joints is therefore proportionately increased ...."

"As an illustration of the effects of increased
pressure in increasing the corroding action, it may be
mentioned that this grooving of the plates has been
found to be materially increased in amount since the
working pressure of locomotive has been increased from
100 lbs up to the present 140 lbs/in^2."

"In the case of the boilers constructed in the
ordinary manner (lap), the plates cut out show the
grooving action of the corrosion below the water line,
while they are comparatively clean above."

From these observations, Kirtley concluded that
the corrosion of the plates at the joints was to be
attributed to the chemical action of the water in the
boiler being concentrated upon those particular parts
in consequence of the mechanical stress. He was able
to appreciate this because he observed that the middle
plates which were relatively free from stress and only
subject to chemical action showed no further furrowing.
Joints in the upper part of the boiler above the water
line were also free from furrowing, being exposed to
the mechanical action alone, without the chemical action.

Tyler [17] and later Kirtley [19] certainly had a
reasonable understanding of the problem of furrowing.
They failed, however, to appreciate fully that the
crack itself formed a resting place on which deposits
could settle. Tyler considered that the resting place
had been removed with the change of construction. At

this stage in the development of corrosion science,
there was little appreciation of the problems of stress-
corrosion and corrosion fatigue. A point of significance
was that the problem was partly understood and logical
attempts within the confines of their experience were
being made to overcome it.

    Within a few months of Tyler's paper, the 'Western
Times' of Exeter reported another boiler explosion
in which:-
".... The plates near an angle-iron junction had
'furrowed' or become corroded in a deep groove, whereby
the thickness had been reduced to ^{1}/16 inch and finally
gave way making a rent 20 x ¼ inch wide. No other
damage was observed, 30 people were killed."

    However, the 'Manchester Boiler Association' [21]
reported that they had examined 458 boilers of which
47 were badly corroded and 5 very dangerous. No accidents
had occurred to boilers inspected by the Association,
although many others had exploded.

    The following year, the incidence of boiler
explosions continued. A report in 'The Engineer' stated:-
[22]
".... There have been a greater number of boiler explosions
this year than ever before. Usually about 25 explosions
occur yearly in Great Britain. This year already 40
have occurred. The use of steam power is increasing
and while many thousands of boilers are growing old,
higher pressures are being used." Fig 1 shows the
inspection following an explosion at Blackburn in 1874.

    As with modern industry, the demands for greater
performance continued requiring that boilers performed
at greater pressures and under more exacting circumstances.

In conclusion

It was only through the growth of organisations such as
the Manchester Boiler Association and the attitudes of
more enlightened managers that a greater appreciation
of the problems involved became apparent.

REFERENCES

1   W Austin, Phil. Trans., 1788, 78, 379.
2   W Nicholson & A Carlisle, Nicholson's J., 1800,
      4, 179.

3   H Davy, Nicholson's J., 1800, 4, 337.
4   W H Wollaston, Phil. Mag., 1801, 11, 206.
5   Anon., Ann. Chem. Phys., 1819, 11, 40.
6   M Hall, Quart. J.Sc., 1825, 25, 262.
7   H Davy, Phil. Trans., 1826, 116, 383.
8   A C Becquerel, Ann. Chim. Phys., 1827, 35, 113.
9   R Mallet, British Assoc. Rep., 1840, 2, 221.
10   T Andrews, Proc. Roy. Soc., 1884, 37, 28.
11   T Andrews, Trans. Fed. Inst. Min. Eng., 1890,
       -, 191.
12   R Adie, Phil. Mag., 1847, 31, 351.
13   W R Whitney, J. Am. Chem. Soc., 1903, 25, 395.
14   T Nernst, Proc. Inst. Civ. Eng., 1889, 4, 150.
15   Anon., 'The Engineer', 1862, 14, 293.
16   W H Tyler, 'The Engineer', 1863, 15, 311.
17   E B Marten, 'Boiler Explosions', E & F N Spon 1878.
18   W Kirtley, Proc. Inst. Mech. Eng., 1866, 17, 60.
19   Anon., 'Western Times', 1862, 25th April.
20   Anon., 'The Engineer', 1862, 14, 261.
21   Anon., 'The Engineer', 1863, 15, 295.

Priestley Lecture 1989

A Scientist's View of Religion

Reverend Dr John Polkinghorne, FRS

PRESIDENT OF QUEENS' COLLEGE, CAMBRIDGE, UK

It is an honour for me, which I greatly appreciate, to give this year's Priestley lecture. I am also grateful for the ecumenical gesture of inviting a physicist to address an audience of chemists.

Joseph Priestley was a man of serene temperament, insatiable intellectual curiosity, and great clarity of mind. He was, perhaps, the most distinguished product of the eighteenth century dissenting academies, those remarkable educational institutions which more than made up for the barring of non-members of the Church of England from the Universities of Oxford and Cambridge. There is much to admire in Priestley but, like most of us, he had the defects of his qualities. They were nowhere more evident than in his religious opinions. He is associated with a form of belief which is usually called Rational Dissent. Its characteristics are a certain narrowness of vision but possibility and an undue reliance on what can be considered a *priori* reasonable. The history of science since Priestley's day provides us with many examples of how strange and unexpected the-way-things-actually-are turns out to be: how limited are our powers of intellectual prevision, which often stand in need of what David Park has called 'universe-assisted logic' - that is, the prompting of our recognition of new and beautiful forms of rational structure which would not have occurred to our unaided imagination.

Priestley's Enlightenment-encouraged confidence in the power of reason, narrowly construed, led him to reject much of traditional Christianity. He failed to recognise that the mysterious but exciting doctrines of the incarnation (that God has made himself known in the plainest possible way by living the life of a man in Christ) and of the Trinity (that God is complex and superpersonal in a way that transcends human experience of isolated individuality) arose, not as rash and ungrounded metaphysical speculations, but as the result of centuries of struggle to come to terms with what the Church believed to be its essential and undeniable experience.[1] Theological doctrines are as much the consequence of reflection upon experience, and the search for its adequate understanding, as are the theories of physical science.

But - you may well say - how much more successful have been the latter than the former! Since Priestley's day we have learnt, and can agree that we have learnt, much about the gases that he discovered, even to the level of seeing them as made up of quarks and gluons and electrons. Yet in the religious sphere there has been no comparable growth of agreed understanding. Rather the reverse, for the varieties of belief and unbelief have greatly multiplied since the eighteenth century, to the point where Priestley's dry unitarianism, at the time considered dangerously radical, might be held by a neutral observer to occupy a place much closer to the centre of the spectrum of possible opinion. And isn't it just that last word - opinion - that sums it all up? Science deals with a public world of fact and trades in real knowledge. Religion is concerned with a private world of opinion, where we are all entitled to our personal view but these are incommensurable and nothing can adjudicate between them.

I want to try to explain why I emphatically reject that option for the peaceful coexistence of a universal science with an individual religion. I believe my concern as a physicist and my concern as a theologian are both with the search for truth - not just 'truth for me' but 'truth' *tout court*. My scientist's view of religion is that it is the exploration of aspects of reality complementary to those which we encounter in our scientific investigations. In fact, religion is the encounter with Reality (with a capital R), the One who is the ground of all else that is. The substantiation of that claim demands much argument, some of which I shall attempt to sketch within the confines of this lecture.

But first, I must try to say something about why
theology does not display clearly that cumulative and
progressive character which is so impressive a feature
of science.   At root the answer is that science is
easy.   Most of you will probably give a wry smile at
that, conscious of the formidable difficulties that
your own investigations have to overcome.   Of course I
know that too, but what I mean is that in our
exploration of the physical world we enjoy some
formidable advantages. We transcend that world, in the
sense that we can interrogate it and put it to the
test.   We have the weapon of experiment, a weapon whose
value I, a *theoretical* physicist, gladly attest.   There
are many realms of human experience in which we loose
that probing power.   It is so in all forms of personal
encounter.   If I am always setting little traps to see
if you are my friend, by the very lack of trust that I
display I shall destroy the possibility of true
friendship between us.   Even more so is that true of
the encounter with God.   It is no use saying 'if there
is a God let him strike me down dead'; he doesn't play
that sort of silly game.   In the realm of the personal
the ability to test is replaced by the necessity to
trust.

Impersonal knowledge, such as that which is the
staple of science, is lowest-common-denominator
knowledge. It is concerned with general features of the
situation which can be present time and again and it is
able to dispense with the consideration of many idio-
syncrasies, such as the colour of the experimenter's
hair or the state of his digestion.   This essential
repeatability permits predictability, one of the most
impressive features of science's success. On the other
hand, every personal encounter is unique. We never hear
a Beethoven quartet the same way twice, even if we play
the same record.   Only in the event itself is its
meaning to be found, nor can any casual bystander say
what it is.   Someone who seeks wholeness in the
experience of serious illness may find it in the
restoration of health or in the acceptance of the
imminent destiny of death.   No one can say beforehand
which it shall be and only the person involved can say
whether it has been found.

Impersonal science makes great strides by taking
things apart but in personal encounter we must accept
things in their totality.   There must be many people
here who could give in chemical terms an exhaustive
account of a Rembrandt self-portrait, specifying the

composition of every speck of paint, but I am sure that there is no one here so foolish as to suppose that a Rembrandt self-portrait is just a collection of specks of paint.

If Einstein had not discovered the General Theory of Relativity, then no doubt, in due course and perhaps piecemeal, others would have hit on the possibility of a gravitational theory framed in geometrical terms. No one, not even the greatest genius, is indispensable to science. Personal knowledge, however, cannot escape from the scandal of particularity, the irreplaceability of the individual. Only Bach could have written the *Mass in B Minor*. In the sphere of religion it is not inconceivable that the fullest knowledge of God was in the possession of a wandering carpenter in a peripheral province of the Roman Empire, far away and long ago.

These contrasts between impersonal science and personal religion, between testing and trusting, predictability and idiosyncrasy, generality and particularity, help me to understand why theology cannot enjoy the degree of collective and cumulative agreement achieved by science. The contrast between the two is not absolute, however. There are the great religious traditions which exemplify a shared and developing understanding of the divine, but their irreducibly personal character means that they can only be evaluated from the inside by participation, and not from the outside by dispassionate inspection.

Let me emphasise that the contrast is between the impersonal and the personal, not between rational inquiry on the one hand and the stubborn and incorrigible fideism on the other. The religious man is not someone who believes twelve impossible things before breakfast. It is as natural for him as for the scientist - and, of course, for many of us the two are the same person - to ask the question What is the evidence? What makes you think that this might be the case? The difference lies in the kind of evidence which is relevant and the consequent manner in which we engage with it. It is also the case that the kind of questions that are addressed point in somewhat different directions. It is a commonplace to suggest that science asks the question How? and is concerned with process; religion asks the question Why? and is concerned with purpose. There is some truth in that, though the two questions are not as separable as one might at first suppose. The tone of theological discourse has certainly been affected by the realisation that humankind did not appear ready-made some few thousand years ago but has emerged as the result of a long and subtle evolutionary process. To suppose that our perception of human worth must be modified by the insights of evolutionary biology would be to embrace what philosophers call the genetic fallacy, but we certainly think differently about the world as the result of Darwin's great discovery.

In fact, if I am right in my view of religion as being as much embedded in experience, and as much concerned with the quest for understanding, as is science in its own domain, then the two must have some degree of impact upon each other. For convenience we may talk about personal or impersonal encounter with the world that we inhabit, but it is still that one world about which we are speaking. Science and religion view it from greatly differing perspectives but their accounts cannot be totally independent of each other and where there is an intersection we must seek consonance between what each has to say in its own appropriate idiom. I want to devote the rest of this lecture to the brief exploration of seven points of contact between the world views of science and of Christian theology.

1. The first thing to note is that science is possible; that we can understand the world. We have benefitted so much from this that we tend to take it for granted. In my view it is a non-trivial - a mathematician's word for highly significant - fact about the world. I am principally referring to what Eugene Wigner, in a famous phrase, called 'the unreasonable effectiveness of mathematics'. Time and again it has proved to be the case in fundamental physical science that the successful theories are those which are economic and elegant in their formulation, which, in a word, are mathematically beautiful. So much is this so that if you want to upset one of your friends working in, say elementary particle physics, you have only to say to him or her 'That latest theory of yours looks a little ugly and contrived to me'. You will be asserting that it lacks the character, so highly prized by people like Paul Dirac and Erwin Schrödinger, and repeatedly found to be present in empirically successful conclusions, of yielding beautiful equations with an almost self-authenticating character to them, making one think 'That must be right'.

When we use mathematics in this way, as the key to unlock the secrets of the physical universe, something odd is happening. After all, the beautiful patterns of mathematics are mental artefacts, dreamed up by our pure mathematical friends in the privacy of their studies. What I am saying is that some of the most beautiful of these patterns are actually found to be instantiated in the structure of the physical world around us. In other words, the reason within (mathematics) and the reason without (fundamental physics) appear to be in perfect consonance with each other. That is an intriguing fact. It even puzzled Einstein, who said 'the only incomprehensible thing about the universe is that it is comprehensible'.

Now, it is always possible to shrug one's shoulders and say 'Well, that's the way it is, and good luck to you chaps who happen to be smart at maths'. My instincts as a scientist are not to be content with that. After all, it is the desire to understand as fully as possible which motivates our scientific labours. That quest is not to be given up prematurely. Yet physics is powerless to explain its own founding faith in the mathematical intelligibility of the world. If we are to make sense of that we must look elsewhere.

Easy, you say. Evolutionary biology will do it for us. If there were no consonance between the workings of our minds and the way things are we would have perished long ago in the struggle for life. That must be true, but only up to a point. What counts for survival is everyday experience (the world of rocks and trees) and everyday thought (at the most mathematically, arithmetic and the geometry of Euclid). For sure, if those did not match, we would not be here. But I am not talking at that pedestrian level. I am thinking of way-out physical experience, such as the counterintuitive behaviour of the quantum world, and highly abstract and sophisticated mathematics, such as gauge field theories. I cannot believe that quantum field theory is a spin-off from the evolutionary struggle to survive.

The natural way to explain why the reason within and the reason without fit together so perfectly would be if they had a common origin in a deeper rationality. Theism is able to provide just such an explanation, for it sees the rational will of the Creator as the sustaining ground of all our experience, both physical and mental. The universe is rationally transparent,

and beautiful in its transparency, because there is a
Mind behind it.

I do not present that as a knockdown argument. In
the realm of fundamental metaphysical belief there are
no such logically coercive arguments at the disposal of
anyone, believer or unbeliever alike. I do present it
as a coherent rational possibility, made more economic
and attractive for me because I believe that there are
many other grounds for believing that there is a God
who is the Creator of the world.

I   also   proffer   that   insight   as   a   typical
illustration of the way that theology and science
relate to each other. I do not think that in its own
domain science needs augmentation from theology. We
have every reason to think that scientifically poseable
questions will prove to be scientifically answerable,
however difficult it may sometimes be to find those
answers. To believe the contrary would be to invoke
that pseudo-deity, the god of the gaps, popping up as
the 'explanation' of the currently inexplicable and
ever in danger of vanishing with the next advance of
knowledge. If there is a God he cannot just be
associated with the intellectually murky bits of the
universe. He is to be found, not within science, but
beyond it. From our scientific investigations there
arise questions which are not scientific in character,
and so are not scientifically answerable, but which
nevertheless insistently demand a response. The
intelligibility of the physical world is an example.
These metascientific questions point us, I believe, in
a theological direction. For me theology provides the
most profound and comprehensive setting within which to
pursue the search for total understanding. Such a view
of theology's role has been a continuing tradition, at
least since the time of Thomas Aquinas. A distinguished
Thomist philosopher-theologian of this century, Bernard
Lonergan, put the matter in lapidary form. He wrote
'God is the all-sufficient explanation; the eternal
rapture glimpsed in every Archimedean cry of Eureka'.

2. Rather similar considerations apply to the next
point of contact between science and theology. It
concerns the Anthropic Principle. I am sure that its
insights are familiar to many of you. Let me summarise
the matter this way. Suppose God lent you the use of
his universe-creating machine. As you approached this
no doubt formidable piece of machinery, you would find
that it was furnished with a row of knobs that you

could adjust in order to specify the universe you wished to bring into being. One would be labelled 'gravity' and by turning it up or down you could increase or diminish the strength of that fundamental force in your world. There would be knobs for the other fundamental forces and, perhaps, for other particular specifiable circumstances. For example, how big would you like your world to be? Like ours, with a hundred billion galaxies each with a hundred billion stars, or just a cosy cosmos the size of the Milky Way? You choose, pull the handle, and out comes your selected universe. You then have to be patient and wait a few billion years to see what will happen. Our understanding is that unless you had finely tuned those knobs to settings very close indeed to those corresponding to a universe like this, a very boring world would result. It would not produce in the course of its history anything so complex and interesting as you and me.

I won't take time to attempt to explain in detail how we reach this fascinating conclusion. The matter is very thoroughly discussed by John Barrow and Frank Tipler in *The Anthropic Cosmological Principle*. I am more concerned with inquiring what we make of the scientific recognition that a world capable of producing men and women is not 'any old world' but one that is 'finely tuned' in its given law and circumstances.

Once again, one can shrug one's shoulders and say 'We're here because we're here, and that's that'. Such intellectual sluggishness does not commend itself to me. Science itself cannot help us, for it does not explain its own law and circumstances but must treat them as its given starting point of explanation. It has been suggested that maybe there are many different universes and if that were indeed the case then it would not be altogether surprising that one, just by chance, was sufficiently well-tuned to produce life – and of course that's the one we live in because we could not turn up anywhere else. This rather prodigal suggestion has sometimes been defended by an appeal to the many-universes interpretation of quantum theory. I am not persuaded by this, both because I am highly sceptical of that interpretation and also because, even if it were correct, the multiplicity of universes to which it refers arises from the varieties of possible outcomes of quantum events and in no way refers to variations in the underlying physical laws, such as

would correspond to altering the strengths of gravity
or electromagnetism.

The 'portfolio or different universes' proposal to
explain this universe's anthropic fruitfulness should
be recognised for what it is: a metaphysical
speculation. Scientifically we have reason alone to
speak of this universe. I am not at all against
metaphysical speculation, acknowledged as such, but I
regard that alternative metaphysical speculation of the
existence of God as being equally coherent, more
economic and more widely supported by other
considerations. Then this universe would indeed not be
'any old world', but a creation endowed by its Creator
with the necessary physical laws and circumstances to
produce the fruitfulness which is in accordance with
this purpose for it.

3.   That fruitfulness manifests itself in the way
that a universe, initially extremely simple and almost
homogeneous, has become, in the course of fifteen
billion years, highly differentiated and, in parts,
highly complex. Yet that process - whether we are
thinking of the fluctuations of matter density which
initiated the condensation of the galaxies, or the
currently unknown biochemical pathways by which amino
acids aggregated to produce the seeds of life, or the
biological story of the evolving complexity of life -
all these developments may be characterised as resulting from the continuing pattern of the interplay of
chance and necessity. Chance (by which I mean the
uncorrelated instances of happenstance) is the source
of novelty. Necessity (by which I mean the reliable
laws of nature) provides the means to sift and preserve
the offerings of novelty. Does not the unforseeable
role of chance - so that what will be cannot be read
out of what is - subvert the religious claim that there
is a purpose at work in the world's process? Jacques
Monod expressed this view with passionate Gallic
rhetoric when he wrote that 'pure chance, absolutely
free but blind, lies at the base of the stupendous
edifice of evolution'. The crucial word here is
'blind'. For Monod the universe is a tale told by an
idiot. In the face of it, the only conceivable stance
is that of heroic defiance.

One does not need to see it that way. In particular
one must not lose sight, as Monod so often seems to do,
of the essential role of necessity in all its anthropic
fruitfulness. Let me put an alternative view in the

following way. With respect, one might feel that God faced a dilemma when he came to create the world. The God of faithfulness will surely create a reliable world, but then there is the danger that its regularity will be so rigid that it is no more than a home for automata. The God of love will surely create a world endowed with freedom, but there is then the danger that the very openness of such a world will degenerate into licence and chaos. The universe that we actually perceive, with its balanced and fruitful interplay of chance and necessity, novelty and regularity, is a world that one might expect as the work of a Creator both loving and faithful, for it incorporates his twin gifts of freedom and reliability.

4. I have spoken of the openness of the universe. In the eighteenth century, following on Newton's great achievements, it seemed as though we lived in a world of mechanism. Joseph Priestley himself was a necessitarian, regarding even human life as the unfolding of an inexorable chain of consequence. I believe such a view to be inconsistent with our basic experience of choice and responsibility. I also believe it to be self-destructive. Rational discourse would be abolished and replaced by the mouthings of automata. In practice, even the most fervent necessitarian exercised a tacit disclaimer on behalf of his own thought and argument. Yet a century which constructed orreries, those mechanical models of the solar system in which the turning of a crank caused the planets to circle with relentless regularity, was almost bound to take a rigid view of physical process.

All that has dissolved away in the twentieth century. Not only has quantum theory revealed our apparently clear and determinate world to be cloudy and fitful at its atomic roots, but also we have become aware of how special and untypical are those predictable systems of planets or pendulums on which we cut our dynamical teeth as students. For sure, there are some clocks around but most of the physical world is made up of clouds. Before such an audience as this I do not need to go into details about the exquisite sensitivity displayed by complex dynamical systems, which makes them intrinsically unpredictable. The modern theory of chaos - that subtly structured randomness - is a development of very great significance.

The picture presented to us is that of a system of underlying equations which are determinate but whose solutions are so precisely and critically sensitive to exact initial conditions as to make them inaccessible to prediction. In other words the immediate and widely agreed consequence is *epistemological*, concerned with what one can know. If one goes on to speak, as I have done, of a genuine openness of the future, then one has moved to making an *ontological* statement, concerned with what is the case. That is a more controversial matter, but one which, I believe, can be defended as being a natural step to take.

I do so on two grounds. The simplest one is that such openness accords with our own experience as human beings and it is a gain to *physics* to be able to describe a world of which we can conceive ourselves as inhabitants. The second ground is more philosophical. Like almost all scientists I am a realist, that is to say I believe that our investigations lead us to reliable knowledge about aspects of the physical world as it actually is. That claim, so natural to a scientist (for why else should we go to all the immense trouble that doing science involves?) has been hotly contested by many philosophers of science. This is not the place to refight that battle; I shall simply assume victory.[2] For a realist, what we can know (epistemology) and what is the case (ontology) are intimately connected. Thus it is wholly natural to move from unpredictability to openness. We have seen it happen before. Heisenberg's analysis of uncertainty principle - the gamma-ray microscope and all that - was epistemological, for it was concerned with what could be measured. Almost all quantum physicists have gone onto embrace an ontological interpretation, asserting an intrinsic indeterminancy in the simultaneous position and momentum of quantum entities.

Therefore I believe the apparently determinate equations of classical physics to be themselves approximations in the description of a more flexible and subtle physical reality. Mere mechanism is dead; the future is not a tautologous spelling out of what was already there in the past, we live in a world of dynamic becoming as well as static being.

I have already associated our room for manoeuvre as persons with aspects of this openness of physical process. I do not think that we exhaust it. There will also be a freedom for the whole universe to explore and realise its own potentiality. And I do not see why God, that universe's Creator, should not have reserved some part of that openness for himself in his interaction with the world. In other words, the providential relationship of God to his creation seems to me to be a coherent possibility. Such interaction would be hidden within the cloudy unpredictability of complex systems. It would not be demonstrable by experiment but would be discernible only by faith.[3]

The picture I am offering seeks to steer a course between two unacceptable theological extremes. It eschews the deistic picture of a God who simply keeps the world in being but who has no particular relation to what is going on within it. Equally, it eschews the picture of God as Cosmic Tyrant, making the whole of his creation jump to his tune alone. That creation, and ourselves within it, enjoys a genuine freedom. God respects that, but not to the point of enforcing his own total impotence.

There are many issues here which need detailed consideration. Let me be content to draw attention to just one. The greatest problem of all for the theist is the problem of evil. I do not need to spell it out for you, for we all know how grievous it is in this post-Holocaust, disaster ridden, world. If God is good and powerful, why does he allow such things to happen?

There are two sorts of evil. One is moral evil, the chosen cruelties of humankind. There are no simple answers to them, but many have felt that some insight is to be had through what is called the free-will defence: that it is better to have a world of freely choosing beings, however disastrous some of their choices may be, than to have a world of perfectly-programmed automata. Our reluctance to countenance coercive measures such as the castration of persistent sex-offenders shows that we feel some force in that argument.

It does nothing to explain physical evil. Clearly illnesses and earthquakes are not caused by humankind. If I am right in thinking that God gives freedom to all his creation (and not just to men and women), then the free-will defence needs augmentation by what one might call the free-process defence. Austin Farrer asked himself what was God's will in the Lisbon earthquake, that great disaster during Joseph Priestley's lifetime which killed fifty thousand people in one day. He concluded that it was that the elements of the Earth's crust should behave in accordance with their nature. It is a hard reply, but I believe is a true one. God does not will the incidence of a cancer anymore than he wills the act of a murderer, but he allows both to happen in the free world that he sustains in being.

5. How will it all end? As you know, taking the broadest and longest view, badly. Science suggests that the universe will follow one of two possible scenarios: either collapse (as the galaxies fall back into the melting pot of the Big Crunch) or decay (as they continue to separate for ever but themselves degenerate into a low energy plasma). Thus it seems that ultimately the universe is condemned to futility. What does that imply for the religious claim that there is a purpose at work in the world?

I think that the answer is that it throws it back upon the only possible source of lasting hope, which is God himself. It has never been part of classical Christian thinking to subscribe to an evolutionary optimism, seeking fulfilment within physical process alone. The problem of the decay of the universe, on a time scale of billions of years, is not all that different to the problem of the decay of ourselves, on a time scale of tens of years. In both cases, if there is a continuing destiny, its basis must be the faithfulness of a loving God. I therefore go on to consider whether the Christian hope of a human destiny beyond death is still credible in a scientific age.

6. Joseph Priestley did not believe in the soul as an independent spiritual component of humanity, capable of surviving the death of the body. In his view, when the brain died the mind died and it remained dead until resurrected at the end of time. It may surprise you to know that my own views are quite close to those of Priestley.

If we are to talk of an individual destiny beyond death, we have first to decide what is the real 'me' that we are talking about. Obviously it is not the atoms making up my body, for they are changing all the time. I personally believe that the real me is the almost infinite information-bearing pattern in which those transient atoms participate at any given time. It is that pattern, in its developing continuity, which is the true meaning of the soul. We are not apprentice angels, spiritual beings temporarily trapped in a

fleshly prison, but psychosomatic unities - as the
Hebrews knew long ago.   At death that pattern is
dissolved but it seems to me to be an entirely coherent
possibility that the pattern is remembered by God and
recreated by him in some unimaginable future environment of his choosing.   In other words, I embrace the
Christian hope of death and resurrection, depending for
its conviction on the faithfulness of God, and not a
hope of survival, depending upon the intrinsic
immortality of a purely spiritual soul.   I do not think
that scientific advance has in any way made that
Christian hope incredible.

While the actual realisation of resurrection lies
for us beyond present time, it is part of the Christian
faith to believe that the eventual human destiny has
been anticipated within history for one man, in the
resurrection of Jesus.   That brings me to my final
issue.

7.   Christianity's assertion of the resurrection of
Jesus as one of its central beliefs raises inescapably
the question of whether miracle (that is, marvellous
and wholly unprecedented occurrence) is still credible
in a scientific age?   After all, we are deeply
impressed by the regularity of the workings of the
world that has been revealed to scientific inquiry.
Joseph Priestley found himself quite unable to accept
the resurrection of Christ.

I am not so sceptical.   It is important to
recognise that the problem of miracle is basically
theological rather than scientific.   By its very
nature, science is badly placed to adjudicate on the
unique and unrepeatable.   The difficulty that theology
faces is that God's relationship with the world must
surely always be characterised by utter consistency. It
is theologically incredible that he should ever act as
a celestial conjurer, today doing a turn to impress
someone or to help a particular friend, while yesterday
he had not thought of it and tomorrow he will not be
bothered to repeat it.   In the face of claims of the
miraculous, theology has to explain how it is that a
particular astonishing happening is consistent with
God's continuing faithfulness; why Jesus was raised
from the dead whilst it is our common experience that
dead men stay dead.

I think that the physical notion of a new regime
provides a helpful analogy.   We all know how underlying

regularity of physical law can be combined with
startling changes in the consequences of that law, for
instance at a phase change from one kind of regime to
another. Similarly, if it is true, as Christians
believe, that God was present in Jesus Christ in a way
in which he has not been present in any other man, then
it is at least a rationally coherent possibility that
that new regime that Jesus represented should be
accompanied by new phenomena. Obviously, very much
more needs to be said before a belief in the
resurrection could be justified. In my view, there is
evidence to which one can point in building a case for
that belief, but it would not be appropriate to pursue
that here.[4]   I simply want to assert that the
possibility is not ruled out from the start, for we do
not possess a knowledge of what may be possible in
unprecedented circumstances. My understanding of the
resurrection (or any other claimed miracle) is that it
is never just a divine *tour de force* of brute power,
but that it is always related consonantly to God's
consistent and continuing action in the world. In
other words, I see miracle as unexpected providence in
unprecedented circumstances.

My lecture is almost at an end. It has been
concerned principally with the border between the
scientific and theological world views. I stand before
you as someone who is both a physicist and a priest and
who seeks to take with equal seriousness the insights
of both kinds of experience. I have tried to sketch how
they may be held together, not without puzzlement at
times, but without dishonesty or compartmentalism and,
indeed, with some degree of mutual benefit and
enlightenment. I have only touched on some aspects of
our scientific understanding. Equally, I have only
touched on some aspects of Christian belief. Central
to my own faith is something I have not spoken of, the
encounter with Christ in scripture, in the Church and
in the sacraments. Let me conclude by giving just one
example of how that central ground of Christian
experience can provide a more profound response to one
of the problems we have looked at together. I refer to
the problem of suffering.

Even if the free-will and free-process defences
offer some insight, they do not operate at levels
sufficiently profound to meet fully the agonising
challenge of evil. We encounter it in the depths of
our being and it demands a more than cerebral response.
It is central to my own religious belief that the

Christian God is not just a spectator of the world's suffering, however benevolent, but that he has also been a participant. In that lonely figure, hanging in the darkness and desolation of Calvary, the Christian sees God himself opening his arms to embrace the suffering of the strange world that he has made, and by that act of acceptance offering the hope of an overcoming. That is an insight that moves me most profoundly.

REFERENCES

1. See *The Way the World is*, Triangle, 1983, ch.9.

2. See *Rochester Roundabout*, Longman, 1989, ch.21.

3. See *Science and Providence,* SPCK, 1989.

4. See Ref. 1, ch.8.

Upgrading of Landfill Gas by Membranes — Process Design and Costs, Comparison with Alternatives

R. Rautenbach and H. E. Ehresmann

INSTITUT FÜR VERFAHRENSTECHNIK, RHEINISCH-WESTFÄLISCHE TECHNISCHE HOCHSCHULE AACHEN, TURMSTRASSE 46, D-5100 AACHEN, FRG

1. INTRODUCTION

In the digesters of sewage treatment plants and on landfill depositories, biogas is produced in large quantities. Biogas contains about 50 – 70 vol–% of methane, carbon dioxide, small amounts of water vapor, air, traces of hydrogen sulfide and in the case of landfill gas some halogenated hydrocarbons.

Although biogas is an interesting energy source, its specific energy content is too low for a transport by pipeline even over short distances. There are two alternatives for an economical energy export:

- conversion into power by gas engines / generators on the site and feeding into the grid system /1/
- methane enrichment on the site to about 90 – 96 vol–% and feeding the product into the natural gas distribution system /2, 3/.

The investigations reported here are related to the second alternative and confined to gas permeation using polymer membranes of the solution–diffusion type.

2. MATERIAL TRANSPORT IN GAS PERMEATION

In principle, gas mixtures can be fractionated by porous membranes and by pore–free membranes of the solution–diffusion type. The selectivity of porous membranes is, however, low compared to the selectivities which can be achieved with solution–diffusion type membranes. Accordingly, gas permeation gained interest after the introduction of pore–free polymer membranes, and even then only when they became available as asymmetric, high–flux membranes. The module types employed in gas permeation are of the hollow fibre and of the spiral–wound configuration.

The essential transport steps in such membranes are:

- sorption of the permeating component in the polymer
- diffusion of the permeating component through the polymer
- desorption of the component out of the permeate side.

The material transport of a permeating component is described with suff-icient accuracy – at least for permanent gases like oxygen, nitrogen, and methane – by:

$$\dot{n}_k'' = Q_k \, (p_F \, x_k - p_P \, y_k) = Q_k \, p_F \, (x_k - \delta^{-1} \, y_k). \qquad (1)$$

According to eq. 1, the molar flux $\dot{n}_k''$ of any permeating component k is proportional to the difference of its partial pressures across the mem-brane. The constant Q_k of eq. 1, the permeability of the membrane for the component k, can be determined from permeation experiments with the membrane.

With the assumption of "unhindered" flow, i.e. negligible influence of the permeate concentration on local permeate production, the permeate composition $y = y_i$ of a binary mixture (k = i,j) follows from eq. 1:

$$y = \frac{1 + \delta \, (x + \frac{1}{\alpha - 1})}{2} - \left(\left[\frac{1 + \delta \, (x + \frac{1}{\alpha - 1})}{2} \right]^2 - \frac{\alpha \, \delta \, x}{\alpha - 1} \right)^{0.5} \qquad (2)$$

For design and simulation of membrane modules and processes, two parameters are important:
- the ideal separation factor $\alpha_{ij} \equiv Q_i/Q_j \geq 1$ which follows from the general definition of the selectivity and eq. 1 for $p_P \to 0$
- the pressure ratio $\delta \equiv p_F/p_P \geq 1$.

Figure 1 which is based on eq.2, shows the influence of these parameters on permeate composition y for different feed compositions $x = x_i$. It clearly indicates that for low concentrations of the preferentially permeat-ing component in the feed, a pure permeate cannot be obtained in a one–stage process, even for very selective membranes, since in this case the process is not limited by membrane selectivity but by driving force.

For an ideal membrane ($\alpha \to \infty$), eq.2 yields, for example, y = 0.5 for a feed composition of x = 0.01 and a pressure ratio of δ = 50. This is the maximum achievable permeate composition under these conditions, because the driving force for component i is already zero:

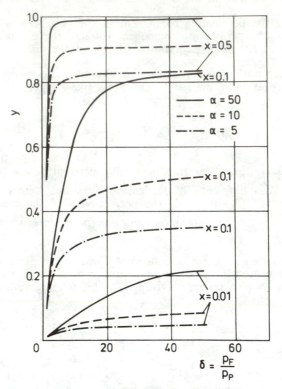

Figure 1
Influence of pressure ratio, ideal separation factor and feed composition
on permeate composition

In general, the driving force for the transport of a component i across
the membrane is the difference of the chemical potentials, $\Delta\mu_i$. For ideal
gases,

$$\Delta\mu_i = R\ T\ (\ln \frac{p_{Fi}}{p_0} - \ln \frac{p_{Pi}}{p_0}) = R\ T\ \ln \frac{x_i\ p_F}{y_i\ p_P} \tag{3}$$

follows directly from the definition of the chemical potential, resulting in:

$$\Delta\mu = 0 \quad \text{for:} \quad y = 0.5, \quad x = 0.01 \text{ and } \delta = 50.$$

It must be emphasized, however, that these considerations concern only
the achievable permeate composition of a one-stage process. In cases like
the production of methane from biogas, where the retentate is the pro–

duct, almost any product purity can be achieved by a one–stage process, however, at the expense of losses of the valuable component.

In principle, the pressure ratio δ can be increased by either raising the feed side pressure or by reducing the permeate side pressure (applying vacuum). The application of vacuum to the permeate side is only feasible in cases where the permeate side pressure losses due to friction are kept very low by adaequate module design. Almost all commercially available gas permeation modules are designed for operation at ambient pressure at the permeate side.

It must be emphasized that eq. 1 has to be applied locally. As a result of flux across the membrane, the relevant parameters like flow rates at the feed and permeate side of the membrane, gas compositions and pressures will vary along the membrane. This has to be taken into account in all module and process calculations.

At this point, a short concideration on module flow pattern seems to be appropriate. Depending on module design, feed and permeate flow can be

- cocurrent
- counter–current
- cross flow.

Hollow fiber modules for gas permeation, for example, realize either cocurrent or counter–current flow, whereas the flow pattern in a spiral wound module is strictly cross flow. The influence of flow pattern on module performance has been discussed in detail elsewhere, for example by Hwang and Kammermeyer /4/. **Figure 2** shows results of a mathematical modeling of air fractionation. In this diagram, the molar fraction of the preferentially permeating component oxygen is plotted against the cut rate for the different flow patterns. According to **figure 2**, counter–current flow gives the best results. Nevertheless, the effect of flow patterns should not be overestimated. **Figure 2** indicates, that there are only marginal differences between counter–current flow and the model "plug flow – unhindered permeate flow" for the chosen conditions. In reality, the differences become even smaller for two reasons:

- the membranes relevant in gas permeation consist of a very thin active layer strengthened by a porous support. This porous support layer acts as a diffusion barrier, diminishing the influence of the permeate composition on local permeate production
- even neglecting this influence, the differences between all flow patterns decrease with increasing membrane selectivity.

In reality, therefore, module design is relatively free with respect to module flow pattern. Furthermore, the relatively simple model assuming plug flow at the feed side and no concentration influence of the permeate side on the composition of the locally produced permeate (unhindered permeate flow) is sufficiently accurate.

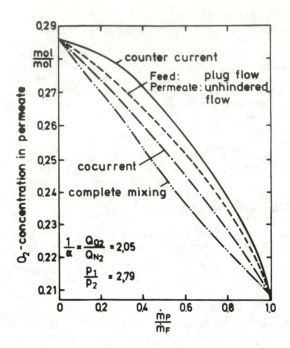

Figure 2
Effect of flow pattern on module performance in gas permeation /4/

3. EXPERIMENTS

Since 1985 the institute operates a pilot plant for methane enrichment of biogas by gas permeation on the premises of a municipal sewage treatment plant. The flexible plant design with feed flow rates up to 70 $m^3(STP)/h$, feed pressures up to 60 bars and maximum temperatures of 130 °C allows the installation of different module systems (**Figures 3, 4**). Three systems, Envirogenics spiral–wound module equipped with cellulose acetate membranes and the hollow fibre modules of Monsanto Co. and UBE Ind. with polysulfone/silicon resp. polyimide membranes have been tested for several thousand hours each. Based on the results of this project, a gaspermeation plant with a capacity of 200 $m^3(STP)/h$ has been constructed and started operation in July 89 on the premises of a landfill depository (**figure 5**). In order to meet pipeline gas specifications, this plant includes a pretreatment stage for the removal of the trace components H_2S and halogenated hydrocarbons (product: H_2S < 1 ppm, Cl,F < 5 ppm).

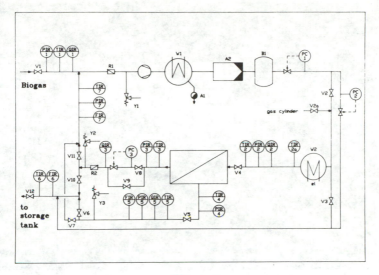

Figure 3
Flow sheet of the gas permeation pilot plant

Figure 4
View of the 70 m³(STP)/h gas permeation plant

Figure 5
View of the 200 m³(STP)/h gas permeation plant

Influence of Operating Parameters on the Process

With gas permeation, the separation effect occurs due to the differences in transport velocity of the individual gas components through the membrane, commonly referred to as "slow" and "fast" gases. In general, all polymers are qualitatively similar with respect to selectivity: they are highly permeable for the biogas components CO_2, H_2O, H_2S and O_2 and less permeable for nitrogen and methane (**figure 6**). Consequently, the product of sewage or landfill gas, the methane, is obtained at the retentate side as mentioned before. It can be deducted from **figure 6** that the maximum achievable methane content is limited by the nitrogen concentration of the raw gas.

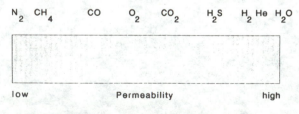

Permeability of Polysulfone / Silicon – Membranes

Figure 6
Relative transport velocity of various gases in polymers /5/

In the following, the operating behaviour of a gas permeation plant for methane enrichment will be discussed. Of major interest is the influence of process parameters such as

- feed gas flow rate
- feed gas pressure
- feed gas temperature

on product composition and recovery rate. If not specified, the standard feed gas composition for the experiments was appr. 67 vol-% CH_4, 32 vol-% CO_2 and 1 vol-% air; water vapor had been removed from the gas by a pressure swing adsorption dryer after compression to a dew point at operating pressure of about –25 °C. The maximum possible transmembrane pressure difference for all three module systems is about 50 bars. In our experiments the feed pressure has been varied between 16 and 40 bars. Pressure at the permeate side (module exit) was kept at atmospheric conditions. Limitations regarding module temperature are de–

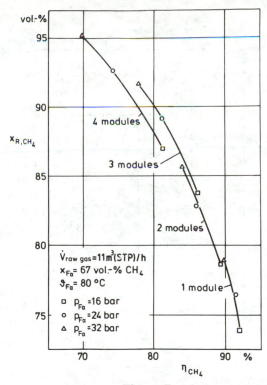

Figure 7
Influence of feed pressure and membrane surface area on product purity and recovery rate

fined by the membrane material with 45 °C for the Envirogenics module, 80 °C for the Monsanto module and 120 °C for the module of UBE IND. Besides the influence of the above mentioned process parameters on product quality and recovery rate, the influence of different module arrangements like one and two stage cascades has been investigated. The Monsanto unit consisted of four and the Envirogenics modules of 15 elements which could be operated in various configurations.

Figure 7 shows the influence of operating pressure (feed pressure) and membrane area on product purity and recovery rate. According to **figure 7**, an increase of the pressure ratio p_{F_α} / p_{P_ω} results in a higher purity of the slower permeating component methane with the disadvantage of reduced recovery rates. With increasing pressure ratio, the feed gas is faster depleted of carbon dioxide. Consequently, the methane partial pressure increases which in turn leads to higher methane flux in the last section of the membrane stage.

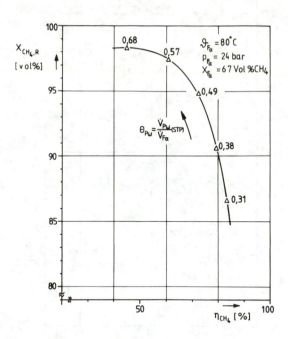

Figure 8
Influence of cut rate on product purity and recovery rate

The influence of the cut rate $\Theta_{P\omega} = \dot{V}_{P\omega}(STP)/\dot{V}_{F\alpha}(STP)$ on methane purity and recovery rate is shown in **figure 8**. In our experiments, the cut rate has been varied by altering the feed flow rate. According to **figure 8**, a further increase in product purity is impossible for cut rates above $\Theta = 0.68$, even if high losses of methane are tolerated. An analysis of the gas composition shows that here the product is almost totally depleted of carbon dioxide, but still contains almost all the nitrogen of the feed. Because of its similarity to methane, it is hardly possible to remove this component by gas permeation.

The influence of feed gas temperature on methane purity and recovery rate is shown in **figure 9**. For a constant feed gas pressure of 32 bars and flow rates of 10 and 20 m³(STP)/h, temperature has been varied between 30 ⁰C and 80 ⁰C. According to **figure 9** an increase in feed gas temperature leads to higher methane purities at the cost of reduced

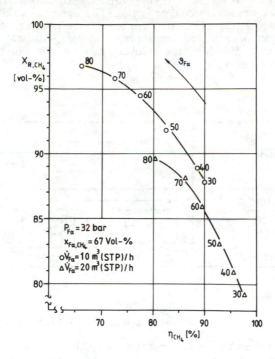

Figure 9
Influence of feed gas temperature on product purity and recovery rate

methane yield. Operation temperature of the modules is important, since generally the flux increases with temperature obeying an Arrhenius type law, while selectivity decreases. It has to be emphasized that **figure 9** shows the influence of the feed entrance temperature on
unit performance and not the influence of local temperature on the local separation characteristic. Because of the integral character of **figure 9**, temperature drops resulting from the Joule–Thomson effect are implicitly contained in these curves. Carbon dioxide and methane exhibit a pronounced Joule–Thomson effect, i.e. a temperature drop in case of isenthalpic expansion. This effect was unimportant in the early applications of gas permeation (hydrogen / nitrogen separation), but has to be taken into account in case of biogas separation.

4. MODULE CONFIGURATION

All results discussed so far have been obtained with a one stage process where all modules were connected in series. The experiments demonstrate that in cases, where the retentate represents the product, any desired product purity can be achieved for a binary system, but at the expense of the recovery rate. However, the recovery rate can be raised by a reflux cascade as shown in **figure 10**.

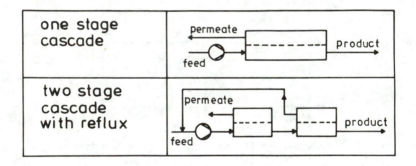

Figure 10
One and two stage cascades for methane enrichment from biogas

Here, an additional compressor is not necessary, but the flow rate is increased compared to the one stage operation. Some results of our experiments are shown in **figure 11**, where a one stage and a two stage reflux process of identical total membrane surface area are compared. According to **figure 11**, the two stage reflux cascade shows significantly higher methane recovery rates.

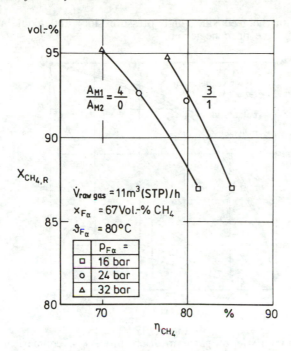

Figure 11
One and two stage cascade with identical total membrane surface area

5. MATHEMATICAL MODELING OF THE PROCESS

Gas permeation is one of the membrane processes where a detailed mathematical modeling is possible and reasonable, since such difficult effects like gel–layer formation, fouling or even concentration polarisation usually do not exist. Based on eq.1 for the local material transport and on permeation experiments with one module element and gas mixtures of different compositions, we developed a mathematical model /6/ of the process accounting for:

- the flow pattern in the individual module
- the multicomponent nature of the gas mixture
- friction losses at the permeate side of the membranes
- the Joule–Thomson effect.

Friction losses at the feed side (high pressure side) and axial backmixing due to diffusion have been neglected.

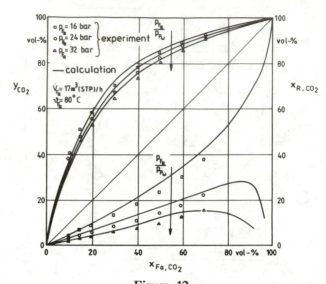

Figure 12

Separation characteristic of gaspermeation process – comparison between experimental and calculated data

In **figure 12**, some results calculated for the one stage process consisting of four modules connected in series are compared with experimental results. According to **figure 12**, the calculations are in good agreement with the experiments. Attention should be given to the fact that the separation characteristic of the unit actually can decrease with increasing pressure ratio as far as the permeate is concerned. This is contrary to the well known local selectivity characteristic which must increase with increasing pressure ratio. The retentate quality, however, increases with increasing pressure ratio, but – as discussed before – at the expense of methane losses.

6. GAS PRETREATMENT – SEPARATION OF TRACE COMPONENTS

The long time operation of our pilot plant has proved that gas permeation with the presently available membranes is a reliable process for upgrading of biogas to the caloric specifications required by the gas distributers. Besides the caloric specifications, however, specifications regarding upper limits of hydrogen sulfide ($\leq$ 1 ppm) and halogenated hydrocarbons ($Cl_2 \leq$ 5 ppm) must be met; these very low figures cannot be achieved by gas permeation. For this reason, our new plant includes a two stage activated carbon adsorption prior to compression (**figure 13**).

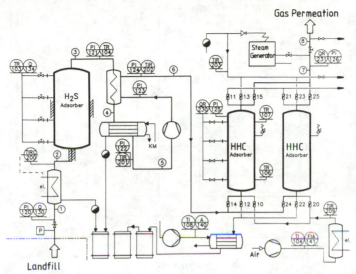

Figure 13
Gas pretreatment for the removal of trace components

In the first stage, hydrogen sulfide is catalytically and partially oxygenated to elemental sulfur which is adsorbed in the pores of the activated hydrocarbon. The sulfur–loaded carbon, containing about 1 kg sulfur per 1 kg carbon, is then deposited or incinerated. The second stage contains activated carbon, specially treated for the adsorption of halogenated hydrocarbons. Other trace components like aromatic hydrocarbons and terpens are also removed in this stage /7/. The second stage is designed as a thermal–swing adsorption, regenerated with steam. The trace components are recovered in form of an aqueous mixture which is further treated by well known conventional methods.

7. COSTS AND PROFITS

Figure 14 is the result of a detailed optimisation calculation for a specific situation. Here, the annual net profit is plotted against the biogas production rate of the landfill $\dot{V}_{biogas}$, which is expected to decrease from currently 480 m³(STP)/h to 360 m³(STP/h or less within 10 years. According to **figure 14**, the optimal strategy is the installation of a one—stage membrane unit (6 : 0) for handling of the present gas capacity, but switching to a two–stage operation by rearrangement of the installed modules later. With decreasing productivity of the landfill, the ratio of

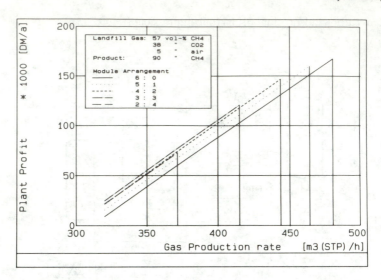

Figure 14
Profit of optimized gas permeation plants for biogas upgrading, influence of landfill gas productivity on module arrangement

membrane surface area of first and second stage $A_{M1}:A_{M2} = N_1:N_2$ should be adapted from $5:1$ to $3:3$ at the most. Even for very low gas production rates of the landfill, ratios smaller than $3:3$ are not advisable. It should be noted that the installed compressor is not affected by these module rearrangements. Feed–side pressure $p_{F\alpha}$ and compressor delivery will remain almost constant (**table 1**).

Naturally, these results are determined to a large extent by the cost of power and the guaranteed price of the product. The calculations are based on contractor bids for the compressors, the membrane modules and the two stage pretreatment. Furthermore, the calculation accounts for operation and maintainance, instrumentation and the necessary installations for a connection to the municipal gas network.

Besides such detailed calculations for a special situation, more general calculations comparing gas permeation (GP) with alternatives /8/ are of interest, i.e. with:

– physical absorption using water (H_2O)
– chemical absorption using monoethanolamin (MEA)
– pressure swing adsorption using carbon molecular sieve (PSA) /9/.

DESIGN DATA

biogas capacity:	480 m³(STP)/h		
module temperature:	50 ºC		
biogas composition:	– methane	57.0	vol–%
	– carbon dioxide	38.0	vol–%
	– nitrogen	4.5	vol–%
	– oxygen	0.5	vol–%
product purity:	– methane	90.0	vol–%

mod.ratio [–]	p_F [bar]	$\dot{V}_{biogas}$	$\dot{V}_{feed}$ [m³(STP)/h]	$\dot{V}_{rec}$	η_{CH4} [%]	Θ_P [–]	profit [DM/a]
6 : 0	40	480	480	–	88.7	0.44	+ 167511
5 : 1	40	464	476	12	90.5	0.43	+ 159786
4 : 2	40	444	472	28	92.4	0.42	+ 146745
3 : 3	40	415	467	52	94.3	0.40	+ 120719
2 : 4	40	372	466	94	96.3	0.39	+ 72722
1 : 5	40	290	491	201	98.3	0.38	– 38696

Table 1
Profit of optimized gas permeation plants for biogas upgrading, influence
of gas production rate on module arrangement

The calculations include the costs for the methane – carbon dioxide separation step, gas treatment for the removal of hydrogen sulfide and halogenated hydrocarbons as well as costs for operation, maintenance and connection to the municipal gas network. The results of a comparison between gas permeation and the alternatives are shown in **figure 15**.
According to **figure 15** where the specific separation costs are plotted against plant capacity, both, psa and gas permeation, are superior to the
absorption processes over the whole investigated capacity range. For
landfill gas production rates over 300 m³(STP)/h gas permeation shows
cost advantages over the psa–process, too. The key data used in these
calculations are listed in **table 2**.

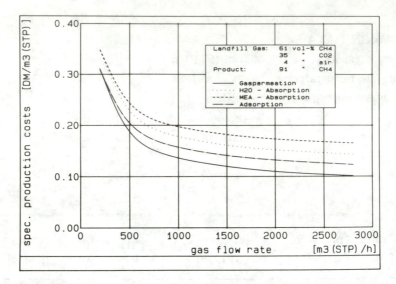

Figure 15
Comparison of specific product cost between gas permeation and
alternative processes

COSTS

compressor	k_c	=	$7689 \cdot P(kW)^{0.697}$	[DM]
electricity	k_{el}	=	0.15	[DM/kWh]
water	k_w	=	0.25	[DM/m³]

DEPRECIATION

interest rate   10 %
period   5 years

Table 2
Cost functions

Since membrane costs are the major part of the total investment costs of a gas permeation, and since gas permeation is a relatively new process, further cost reductions for gas permeation can be expected in the near future.

8. CONCLUSIONS

Biogas, consisting mainly of methane and carbon dioxide, is produced in fairly large quantities (up to 5000 $m^3(STP)/h$) in sewage treatment plants and on landfill depositories. Carbon dioxide can be effectively separated from the biogas by gas permeation, resulting in a product containing 90 % methane or more. Gas permeation is a simple process and, according to our own four years experiments, highly reliable.

In order to meet the specifications of the distributors, certain trace components have to be removed down to very low levels. This requires a gas treatment especially for the separation of hydrogen sulfide and, in case of landfill biogas, the separation of halogenated hydrocarbons, aromatics and terpenes.

The optimal module configuration – "one stage" or "two stage reflux" – has to be evaluated individually since it depends on site–specific factors and on factors like power costs and guaranteed product price.

Compared to alternatives, gas permeation seems to be superior to all absorption processes, independend of capacity. Compared to pressure swing adsorption it is superior at least for capacities of over 300 $m^3(STP)/h$.

NOTATION

Roman Letters

$\dot{n}"$ = permeate flux $[mol/(m^2 \cdot s)]$

p   = pressure   [bar]

Q   = permeability $[mol/(m^2 \cdot h \cdot bar)]$

$\dot{V}$ = flow rate $[m^3(STP)/h]$

x,y   = mole fraction   [–]

Greek Letters

α = ideal separation factor
δ = pressure ratio

μ = chemical potential
η = yield factor
Δ = difference
Θ = cut rate

Subscripts and Superscripts

F   = feed
i,j,k   = gas component
P   = permeate
R   = retentate
α = module entrance
ω = module exit

LITERATURE CITED

/1/   J.Schneider, "Deponiegasaufbereitung und –verwertung im Bereich der Deponie Berlin–Wannsee", Deponie – Ablagerung von Abfällen 2, EF–Verlag, Berlin (1988)

/2/   W.Dahm, H.E.Ehresmann, R.Rautenbach, "Methanerzeugung aus Biogas – Prozessimulation und Betriebserfahrungen", Aachener Membran Kolloquium, 16. – 18. 3. 1987

/3/   H.–J.Schwefer, "Verfahren zur Aufarbeitung von Biogasen in der Gasphase", VDI – Berichte Nr.459, 1982

/4/   S.T.Hwang, K.Kammermeyer, "Membranes in Separation", John Wiley & Sons, New York (1975)

/5/   Company Publication, Monsanto Co., St. Louis, Missouri, 1983

/6/   R.Rautenbach, W.Dahm, "The separation of multicomponent mixtures by gas permeation", Chem.Eng.Process., 19 (1985)

/7/   K.D.Henning, E.Richter, K.Knoblauch, H.Jüntgen, "Reinigung und Weiterverarbeitung von Deponiegas mit Adsorptionsverfahren", Recycling International, 1984

/8/   J.Geuss, "Erzeugung von Erdgas aus Deponiegas", master thesis, RWTH Aachen (1989)

/9/   E.Richter, K.Knoblauch, H.Jüntgen, "Möglichkeiten der adsorptiven Methan–/Kohlendioxidtrennung aus Biogasen", gwf – gas/erdgas, 126 (1985)

Adsorbent Materials: Their Structure, Properties and Characterisation

L. V. C. Rees

PHYSICAL CHEMISTRY LABORATORIES, IMPERIAL COLLEGE, LONDON SW7 2AY, UK

SUMMARY

The properties of zeolites which are important in determining their behaviour as sorbents and molecular sieves are described and compared with the corresponding properties of activated microporous carbons. High silica zeolites and aluminophosphate molecular sieves and their derivatives are included in this comparison. Methods of 'fine-tuning' the effective free diameters of the rings of framework oxygens which control access to the channel network of zeolites are described. Factors which determine the enthalpies of sorption of polar and non-polar sorbate molecules are covered.

The sorption of N_2, CH_4 and CO_2 sorbates in silicalite and NaY zeolites is compared and the factors which control the separation factors of the three binary mixtures of these sorbates in these two sorbents are discussed.

INTRODUCTION

Two very important groups of adsorbents are widely used in the separation of gas mixtures. These are molecular sieves of the zeolite type and microporous, activated carbons.

Zeolites are crystalline aluminosilicates which have very stable frameworks constructed by corner-sharing of all four oxygens of the $[SiO_4]^{4-}$ and $[AlO_4]^{5-}$ tetrahedral primary building units. The crystallographic unit cell of a zeolite is

$$M_{x/n} \, [(AlO_2)_x \, (SiO_2)_y] \, \omega H_2 O$$

where M is a cation of valence η, the ratio y/x can take values from one to infinity and ω is the number of water molecules contained within the pores and channels which permeate the structure. These water molecules can be easily removed by heat and/or vacuum to leave an adsorbent with porosity as large as 50% of the crystal volume. Access to the channel network is controlled by rings of 8, 10, or 12 framework oxygens depending on the zeolite. There are some 40 different zeolites found in nature and over 100 synthetic zeolites. Although some natural zeolites are useful sorbents it is the synthetic zeolites which are widely used to separate gas and liquid mixture on industrial scales.

The cations present in zeolites are there to balance the negative charge introduced into the framework by the substitution of Si^{4+} by Al^{3+}.

Zeolites can be synthesised or by subsequent chemical treatment to have Si/Al ratios which vary from unity to infinity, and the cations can be exchanged by other cations. Thus it is possible to study the sorption properties of specific frameworks with the same surface topology containing sorption sites with electric field gradients which can be varied in density and strength. The effect of these changes on the sorption energies and separation factors of pairs of sorbate molecules which have different degrees of polarity can be easily ascertained.

The first zeolites to be synthesised in commercial quantities (zeolites A, X, and Y) were synthesised from aluminosilicate gels at high pH's and low temperatures containing alkali metal or alkaline earth cations. These zeolites had Si/Al ratios which were very low (1-2.8). Because of their large porosities and high concentration of cationic sites these zeolites had high sorption capacities and large heats of sorption for polar molecules. The synthesis and sorption behaviour of these zeolites are fully described in two excellent monographs by Barrer.[1,2]

In 1972 the zeolite ZSM-5 was synthesised using tetrapropylammonium cations in place of the smaller inorganic cations used up to this time. Because these and similar organic cations are so bulky there is little room in the channels and cavities of the zeolite framework for large numbers of such cations and this limits the number of negatively charged aluminium tetrahedra which can be accommodated in the structure. The net result is the synthesis of a whole new group of high Si/Al ratio zeolites. It was also found that if there was no aluminium present in the starting gel it was possible to synthesise pure porous silica frameworks which did, however, contain some OH defects to balance the positive charge of the cations. By the use of such organic cations it was thus possible to synthesise directly zeolites with Si/Al ratios from 10 to infinity. Many of these zeolites had channels access to which was controlled by rings of ten framework oxygens. Thus these new zeolites were intermediate in their molecular sieving behaviour to the eight ring zeolite A and the twelve ring zeolites X and Y.

Since 1972 a large number of new high silica framework zeolites have been discovered. The synthesis, structure and characterisation of these zeolites is fully described by Jacobs and Martens.[3]

In recent years problems have arisen in the definition of zeolite molecular sieves. An ever-increasing number of frameworks have been synthesised which have channels and cavities of a similar nature to those found in the aluminosilicate zeolites but which contain primary tetrahedral building units of elements other than silicon and aluminium. Some of the high silica zeolites have been synthesised containing small amounts of Fe, B, Ti, etc. in place of, or as partial substitution of the Al in these frameworks. Gallium has replaced aluminium and germanium replaced silicon. All of these materials retain their original aluminosilicate framework structures and can be considered as defect zeolites. The sorptive properties of these materials have not been studied extensively but the substitutions do impart subtle changes to the sorbate/sorbent interaction energies and could lead to improved sorbents for separation processes.

However, a completely new group of molecular sieves have been developed which contain only $[AlO_4]^{5-}$ and $[PO_4]^{3-}$ tetrahedral primary building units in equal numbers. These aluminophosphate frameworks are synthesised hydrothermally at 125 to 200°C under acid conditions (pH $\approx$3). Organic additives (e.g. cationic or neutral amines) are found to be essential to promote crystallization. At least 20 three-dimensional, thermally stable microporous framework structures have been synthesised to date. Fourteen of these ALPO's show zeolitic sorptive properties. Some are zeolite structure analogues but the others are microporous structures with no known zeolite analogue. ALPO-42 has a structure similar to zeolite A while ALPO-37 has the zeolite X, Y structure. Rings of 8, 10 and 12 framework oxygens control the channel size as in zeolites but an aluminophosphate framework containing 18 ring 'windows' has recently been synthesised. This ALPO referred to as VP1-5 has many exciting new sorptive and catalytic possibilities.

Because of the acid conditions which exist during the synthesis it has been found possible to substitute easily the Al and P by many other elements. Thus frameworks containing Al, P and Si have been synthesised (referred to as SAPO's) and frameworks containing Co, Fe, Mg, Mn, Zn, Be, have also been synthesised (referred to as Me APO's).

The net charge on the pure ALPO frameworks is zero. However, the sorptive properties differ from those of the pure, uncharged silica frameworks which are very hydrophobic because of the alternation of Al^{3+} and P^{5+} ions in the framework which does introduce some electric fields into the sorption sites.

An excellent monograph which describes the synthesis, characterisation and sorptive properties of these non-aluminosilicate molecular sieves has recently been published.[4]

Activated microporous carbons are prepared by carbonization and activation of a large number of raw materials of biological origin such as coconut shells, wood, peat and coal. Their structures and properties are related to those of graphite. Riley[5] proposed two types of structures for activated carbons. The first type consists of elementary crystallites that are formed by parallel layers and differ from graphite in that the parallel layers are not perfectly orientated with respect to their common perpendicular axis; the angular displacement of one layer with respect to another being random and the layers overlap one another irregularly. The interlayer spacing of activated carbons is considerably larger than in graphite: 0.344-0.365 nm compared with the 0.335 nm of graphite. The second type is that of a disordered, cross-linked space lattice of carbon hexagons. Some chars display both types of structure.

Selective oxidation of the intermicrocrystalline material and planes of the microcrystallites may develop an extensive porous system according to Rodriguez-Reinoso and Linares-Solano.[6] The resulting pores in activated carbons are classified into three groups (i) macropores > 50 nm diameter, (ii) mesopores between 2 and 50 nm, and (iii) micropores <2 nm. The macropores allow easy access of sorbate molecules to the meso- and micropores. Their 'surface area' is low but they do affect the mass transport properties of the sorbate in the sorbent.

At least 90-95% of the total surface area of the activated carbon is associated with the micropores. These pores of an activated carbon are a little larger than those which exist in zeolites and other framework molecular sieves. The channels in zeolites have constant, accurately defined diameters in the range 0.5 to 0.8 nm while the 18 ring VPl-5 ALPO has channels of 1.2 nm free diameter. Unlike the activated carbons there are no macro- or mesopores in crystalline zeolites as synthesised. However, zeolite frameworks which have been treated chemically (e.g. dealumination treatments) can have amorphous islands in the crystals which show some mesoporosity.

MOLECULAR SIEVES

Access to the channel network of zeolites is controlled by 'windows' of framework oxygens. Zeolites which are important as sorbents and molecular sieves have windows of rings of 8, 10 or 12 oxygen atoms. If these rings are planar the three diameters of these rings are 0.43, 0.60 and 0.77 nm respectively and it is these windows which impart the molecular sieving properties of different zeolites. Examples of zeolites with 8 rings are Zeolite A, Chabazite, Erionite and Rho; with 10 rings are ZSM-5/Silicalite-1, SZM-11/Silicalite-2, Theta-1, EU-1 and Ferrierite; with 12 rings are Zeolites X or Y, Mordenite, Zeolite L and Beta.

The free diameters of these rings can be 'fine tuned' by ion-exchange or pre-sorption of polar molecules. Thus the effective free diameter of the 8 ring in zeolite A is approximately 0.3, 0.4 and 0.5 nm when this zeolite is exchanged into its K^+, Na^+ or Ca^{2+}

cationic form, respectively. In these exchanged
cationic forms the 8 ring contains a K^+ ion, a Na^+ ion
and no cation respectively. These three forms are
usually referred to as Zeolites 3A, 4A and 5A
respectively where the 3, 4 and 5 notation refers to
the approximate free diameter of the window in Å.
Zeolite 3A will sorb water and little else; Zeolite 4A
will sorb oxygen but not nitrogen at 77 K while Zeolite
5A will sorb n-hydrocarbons but not iso-hydrocarbons.

It is also possible to 'fine tune' these zeolite
molecular sieves by presorption of small amounts of
polar molecules, e.g. H_2O, NH_3. These polar molecules
are strongly sorbed on the cations which are sited on
the more open windows of the zeolite and, thus,
increase the effective blocking powers of the cations.

A comprehensive study of the opening of the 8 rings
in zeolite A to the sorption of N_2 at 77 K by ion
exchange of the Na^+ ions with Ca^{2+} ions and the blocking of the same rings to the sorption of O_2 at 77 K by
presorption of NH_3 has been made by Berry and Rees.[7]
In this paper the results obtained were shown to agree
closely with percolation theory which predicts that in
such a cubic array of channels and cages sorption will
occur when 23% of the windows controlling access to
each cage are open.

Separations can be readily effected by use of
differences in the kinetics of sorption of the
components of a mixture. Molecules which are only
slightly smaller than the effective free diameter of
the windows are found to diffuse with diffusion
coefficients some 3 orders of magnitude greater than
those for sorbate molecules which are a 'tight' fit in
the window. Obviously, the fine tuning referred to
above could be exploited to increase such differences.

HEATS OF SORPTION

Differences in the enthalpies of sorption can be used
to separate the components of a mixture. The heat of
sorption, Ψ, is the sum of several interactions:

$$\Psi = \Psi_D + \Psi_P + \Psi_{F\mu} + \Psi_{FQ} + \Psi_{SP}$$

where Ψ_D and Ψ_R are the van der Waals dispersion and
repulsion energies, Ψ_P the polarisation energy, $\Psi_{F\mu}$ are
the energies arising from the interaction of the

sorbate dipole and quadrupole moments respectively with the electric field gradient at the sorption site and Ψ_{SP} is the sorbate : sorbate interaction energy.

Isosteric heats of sorption (q_{st}) are given for various sorbates and sorbents in Table 1.

Table 1 Isosteric heats of sorption (q_{st}) at low surface coverages

Sorbate	NaX	CaX	q_{st}/kJ mol^{-1} NaY	NaA	Silicalite	Activated Carbon
N_2			16.9		7.6	
Ar	11.7	20.9		11.7		
CO_2	51.0		34.4	53.5	23.6	
NH_3	75.0					
CH_4	18.0		18.1		15.3	
C_2H_6	30.5		25.8		28.0	
C_3H_8			34.0		40.0	
C_4H_{10}	55.6				54.0	
Benzene					53.0	48.0
p-Xylene					80.0	

The q_{st} values in Table 1 show the effect of increasing cation density (NaY < NaX < NaA) on the sorption of CO_2. The effect of cationic charge is clearly demonstrated by the sorption of Ar in NaX and CaX. The large q_{st} values which arise from the dipole and quadrupole interactions in the high field gradients which exist in NaX and NaA are seen in the case of NH_3 and CO_2 sorption respectively. q_{st} for CO_2 sorbed in silicalite, where there are no field gradients, is much smaller. The isosteric heats for the hydrocarbons depend only to a minor extent on field gradients and are mainly dependent on dispersion-repulsion forces. The heat of sorption of ethane in NaY is lower than in NaX because of a smaller polarisation energy while the corresponding heat of sorption in silicalite is somewhat larger than in NaY because of the closer fit of the ethane molecules with the channel walls, i.e. a larger dispersion-repulsion energy in silicalite. The heat of sorption of n-hydrocarbons tends to increase by $\pm$ 12 kJ mol^{-1} per CH_2 group. Finally the heats of sorption of benzene in silicalite and activated carbon are similar. The dispersion-repulsion energy will be dominant on these uncharged surfaces and this energy

will be somewhat larger in the more confined channel
network of silicalite compared with the somewhat more
open channels of the activated carbon.   This result
should be repeated with many other sorbates where only
physical sorption occurs but may not be the case for
sorbates such as O_2 where a chemisorption contribution
will occur with the activated carbons.

Although the ALPO's have net neutral frameworks and
contain no cations the sorption behaviour of the ALPO's
differs from that of the corresponding pure silica
molecular sieves because of the differences in the
electronegativity of aluminium and phosphorus.   ALPO's
do exhibit some hydrophilicity.   The intermediate
hydrophilicity of ALPO-5 relative to the very
hydrophilic NaX and the strongly hydrophobic silicalite
can be clearly seen in Figure 1.   The unusual behaviour
of ALPO-5 towards water sorption at different relative
pressures is not fully understood.   There is a need to
study the sorption of various sorbates in some 10 ring
ALPO molecular sieves and compare the isotherms with
the corresponding isotherms in silicalite before the
unusual sorption properties of these ALPO's can be
described.

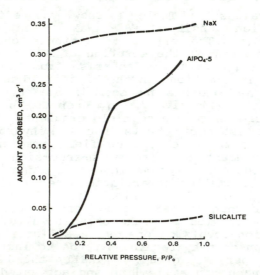

Figure 1   Sorption of H_2O at 24 °C [8]

SORPTION OF N_2, CH_4 AND CO_2 IN SILICALITE AND NaY ZEOLITES

The Henry's Law constants, K_H, have been determined from the initial slopes of the sorption isotherms of N_2, CH_4 and CO_2 in silicalite[9,10] and NaY zeolites[1] and are listed in Tables 2 and 3 respectively. These tables also include the separation factors $\alpha'_{\frac{1}{2}} = K_H(1)/K_H(2)$ for gases (1) and (2) calculated from these constants and the corresponding experimentally determined separation factors $\alpha_{\frac{1}{2}} = (X_1/Y_1)(Y_2/X_2)$ where X_i and Y_i are mole fractions of component i in the sorbed and gas phases respectively. Quite good agreement between the α' and α separation factors can be seen.

The experimentally determined α factors for CO_2/N_2 and CO_2/CH_4 mixtures sorbed in NaY are not given in Table 3. Because of the very strong sorption of CO_2 in NaY at 200 to 270 K the concentration of CO_2 in the gas phase at equilibrium was too small to be measured in these mixtures and the α factors could not, therefore be obtained. These low concentrations of CO_2 are fully consistent with the very large α' separation factors in Table 3 calculated from the K_H constants.

Table 2 Henry's Law constants, K_H and separation factors α' and α for sorption of CO_2, N_2 and CH_4 in silicalite

$K_H/(10^{-6}$ mol kg^{-1} Pa$^{-1})$			α'			α		
Temp./ K								
CO_2	N_2	CH_4	CO_2/N_2	CO_2/CH_4	CH_4N_2	CO_2/N_2	CO_2/CH_4	CH_4/N_2
273	85.0	4.97	17.0	17.04	4.99	3.43	~20	3.5	3.5
283	58.8	4.43	13.4	13.29	4.38	3.04		3.0	2.8
298	35.7	3.78	9.68	9.44	3.69	2.69		2.7	2.7
323	16.9	2.99	6.01	5.65	2.81	2.01		2.5	2.2
343	10.3	2.54	4.31	4.04	2.38	1.69	~ 4	2.0	-

The K_H constant for CO_2 sorbed in NaY at 270 K is approximately fifty times larger than the corresponding K_H constant for this sorbate in silicalite. The difference in the K_H constants for N_2 is only 16% while for CH_4, which is more strongly sorbed in silicalite, K_H is some 75% greater than for NaY. The differences in the K_H constants for these two zeolites are consistent with the differences in the isosteric heats of sorption as given in Table 1 for CO_2 and N_2 but not for CH_4.

Table 3   Henry's Law constants, K_H and separation factors α' and α for sorption of CO_2, N_2 and CH_4 in NaY zeolites

Temp./ K	K_H/(mol kg^{-1} Pa^{-1})			α'			α
	CO_2	N_2	CH_4	CO_2/N_2	CO_2/CH_4	CH_4/N_2	CH_4/N_2
100		2.49	3.60			1.45	
150		2.59×10^{-3}	4.07×10^{-3}			1.57	
200	0.363	8.37×10^{-5}	1.36×10^{-4}	4337	2657	1.63	1.52
247	1.99×10^{-2}	1.50×10^{-5}	2.50×10^{-5}	1327	795	1.67	
270	3.95×10^{-3}	5.79×10^{-6}	9.75×10^{-6}	693	405	1.68	

The separation factors in Table 2 for silicalite indicate that this sorbent would be an excellent sorbent to use in pressure swing separations of mixtures of CO_2, N_2 and CH_4. NaY could also be used to separate CH_4/N_2 mixtures but the very strong interaction of the quadrupole of CO_2 with the large electric fields in NaY although leading to very large separation factors would lead to desorption problems if this zeolite was used in the pressure swing separation of CO_2/CH_4 and CO_2/N_2 mixtures.

REFERENCES

1.   R.M. Barrer, 'Hydrothermal Chemistry of Zeolites', Academic Press, London, 1982.

2.   R.M. Barrer, 'Zeolites and Clay Minerals as Sorbents and Molecular Sieves', Academic Press, London, 1978.

3.   P.A. Jacobs and J.A. Martens, 'Synthesis of High-Silica Aluminosilicate Zeolites', Studies in Surface Science and Catalysis, Elsevier, Amsterdam, Vol.33.

4.   R. Szostak, 'Molecular Sieves: Principles of Synthesis and Identification', Van Nostrand Reinbold, New York, 1989.

5.   H.L. Riley, *Quart. Rev. Chem. Soc.*, 1947, **1**, 59.

6.   F.   Rodriguez-Reinoso   and   A.   Linares-Solano,
     'Microporous   Structure   of   Activated   Carbons   as
     Revealed   by   Adsorption   Methods',   Chemistry   and
     Physics of Carbon, ed. P.A. Thrower, Marcel Dekker,
     New York, Vol. 21, 1989, p.1.

7.   T. Berry and L.V.C. Rees, Soc. of Chem. Ind., Conf.
     on Molecular Sieves, London, 1967, p.149.

8.   S.T. Wilson, B.M. Lok, C.A. Messina, T.R. Cannon
     and E.M. Flanigen, 'Intrazeolite Chemistry', *ACS
     Sym. Ser.*, 1983, **218**, 79.

9.   P.   Graham,   A.D.   Hughes   and   L.V.C.   Rees,   *Gas
     Separation Purif.*, 1989, **3**, 56.

10. P. Graham, A.D. Hughes and L.V.C. Rees,
     in preparation

11. P. Graham, A.D. Hughes and L.V.C. Rees,
     in preparation

Dynamic Modelling of Pressure Swing Adsorption Separation Processes

Douglas M. Ruthven

UNIVERSITY OF NEW BRUNSWICK, PO BOX 4400, FREDERICTON, NB, CANADA E3B 5A3

ABSTRACT

The problem of modelling a PSA separation process has been studied both theoretically and experimentally. A dynamic model based on the linear driving force (LDF) approximation is shown to provide a good prediction of performance for several trace component PSA systems. The comparison between the LDF model and the more complex but realistic pore diffusion model has been examined in detail. While the LDF model can provide a good representation of system behaviour, it is necessary to adjust the proportionality parameter (Ω in $k = \Omega D/R^2$) according to the cycle time. The Glueckauf limit ($\Omega = 15$) is approached only at very long cycle times. The modelling studies have been extended to air separation on a carbon molecular sieve. A good prediction of the performance of a PSA unit is obtained using independently measured kinetic and equilibrium parameters. Results of a recent study of heat effects in PSA systems are also briefly reviewed.

1 INTRODUCTION

Pressure swing adsorption (PSA) separation processes in which the adsorbent bed is regenerated by reducing the pressure and purging, rather than by raising the temperature, were first developed by Skarstrom[1,2] and Domine et al.[3,4]. In its original form the PSA system is best suited to separation and purification processes where it is the less strongly adsorbed species which is required in pure form as the main product. Very high product purities can be obtained but the recovery is relatively low. The process is therefore useful when the feed is cheap so that recovery (or yield) is less important than the production of a high purity product. All three traditional applications of PSA technology (air drying, air separation and hydrogen purification) meet this criterion.

The past decade has seen a concerted effort to extend the range of PSA processes by improving their energy efficiency and by developing process schemes to allow the recovery of either product at high purity. Progress has been reviewed by Wankat[5] and by Yang[6]. Unlike most separation processes which operate at or near the steady state, a PSA system is inherently a transient operation. The interactions between the process variables are complex and difficult to predict intuitively so a suitable mathematical or numerical model is needed to guide design and optimization.

All adsorption separation processes depend on the preferential adsorption of one component (or one family of components) from a mixture but the selectivity may depend either on a difference in adsorption equilibrium or on a difference in adsorption kinetics. All the earlier PSA processes depended on equilibrium selectively. For such systems, the assumption that local equilibrium prevails at all times (equilibrium theory) provides a first approximation to the modelling problem from which useful information concerning the effects of process variables and the conditions required to achieve a pure product, can be derived [Chan et al.[7], Knaebel and Hill[8], Flores-Fernandez and Kenney[9]]. This approach is, however, inappropriate for systems which depend on kinetic selectivity for which a suitable dynamic model is essential.

2 DYNAMIC MODELLING OF TRACE SYSTEMS

The simplest type of PSA system involves the removal of a small concentration of a strongly adsorbed species from a weakly adsorbed or (non-adsorbed) carrier which is present in large excess, so that the fluid velocity may be considered as constant. The heatless air drier and hydrogen purification systems provide practically important examples. We assume isothermal operation according to the Skarstrom cycle (figure 1) with negligible pressure drop through the adsorber bed. The dynamic behaviour of such a system during steps 1 and 3 of the cycle may be described by the following set of equations in which the rate and equilibrium expressions have been written in generalized form:

$$\text{External Fluid:} \quad -D_L \frac{\partial^2 c}{\partial z^2} + \frac{\partial}{\partial z}(vc) + \frac{\partial c}{\partial t} + \left(\frac{1-\varepsilon}{\varepsilon}\right)\frac{\partial q}{\partial t} = 0 \quad (1)$$

For a trace system v is constant so $\partial(vc)/\partial z = v\partial c/\partial z$.

$$\text{Mass Transfer Rate:} \quad \frac{\partial \bar{q}}{\partial t} = f(q,c) \quad (2)$$

$$\text{Equilibrium:} \quad q^* = g(c) \quad (3)$$

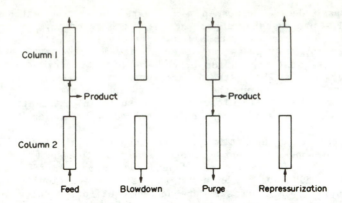

<u>Figure 1</u>   The Skarstrom PSA Cycle

with boundary and initial conditions:

$$D_L \frac{\partial^2 c}{\partial z^2} \bigg|_{z=0} = - v(c\big|_{z=0^-} - c\big|_{z=0^+}) \tag{4}$$

$$\frac{\partial c}{\partial z}\bigg|_{z=L} = 0 \tag{5}$$

$$c(z,0) = h_1(z) \quad ; \quad q(z=0) = h_2(z) \tag{6}$$

$f(q,c)$ and $g(c)$ represent any convenient mathematical forms for the rate equation and equilibrium isotherm while $h_1(z)$ and $h_2(z)$ represent the initial concentration profiles in the fluid and solid phases.

Pressurization and blowdown are rapid so, in a kinetic separation, where the mass transfer rates are relatively slow, as a first approximation, the adsorbed phase concentration profile may be assumed to be frozen during these steps. During pressurization we assume that the feed gas moves in plug flow compressing the residual low pressure gas to the far end of the bed, so that at the end of pressurization we have a step change in concentration between feed and residual gas at a distance corresponding to the pressure ratio. The blowdown (step 2 for bed 2, step 4 for bed 1) is described by the following equations:

External Fluid: $\quad -\frac{\partial (vc)}{\partial z} + \frac{\partial c}{\partial t} = 0 \tag{7}$

Continuity: $\dfrac{\partial v}{\partial z} = \dfrac{1}{P} \cdot \dfrac{\partial P}{\partial t}$ $\hspace{3cm}$ (8)

with the boundary conditions:

$$\dfrac{\partial c}{\partial z}\Big|_{z=0} = 0 \quad ; \quad v\Big|_{z=0} = 0 \hspace{2cm} (9)$$

Any convenient equilibrium expression may be used but in most cases the choice has been either a linear isotherm ($q* = Kc$) or a Langmuir isotherm:

$$\dfrac{q*}{q_s} = \dfrac{bc}{1+bc} \hspace{4cm} (10)$$

The simplest choice for the rate expression is the linear driving force (LDF) approximation:

$$\dfrac{\partial \bar{q}}{\partial t} = k(q* - \bar{q}) \hspace{4cm} (11)$$

which may be used to represent a diffusion controlled process by setting $k = \Omega D_e/R^2$. According to Glueckauf[10,11] $\Omega = 15$ but more recent work by Nakao and Suzuki[12] showed that this approximation is valid only when the cycle time is long compared with the diffusion time (R^2/D_e). The appropriate choice of Ω is discussed in greater detail below.

Since mass transfer is generally controlled by intraparticle diffusion it would be more realistic to use a diffusion equation to represent the mass transfer rate:

$$\dfrac{\partial q}{\partial t} + \varepsilon_p \dfrac{\partial c}{\partial t} = \varepsilon_p D_p \left[\dfrac{\partial^2 c}{\partial r^2} + \dfrac{2}{r}\dfrac{\partial c}{\partial r}\right] \hspace{2cm} (12)$$

$$\dfrac{\partial c}{\partial r}\Big|_{r=0} = 0 \hspace{4cm} (13)$$

$$q(r,t) = g[c(r,t)] \hspace{4cm} (14)$$

$$\bar{q} = 3 \int_o^R \dfrac{r^2 q\, dr}{R^3} \hspace{4cm} (15)$$

To simulate a PSA process the set of equations 1-9 is solved repeatedly in a cyclic manner using the final concentration profile from each step as the initial condition for the next step. Eventually the profiles converge towards the cyclic steady state. Since the repeated solution of the equations is necessary an efficient numerical routine is essential, otherwise the computing time becomes prohibitive. In our studies we have used collocation methods to reduce the partial differential equation to a set of o.d.e.s. which are then integrated by a standard integration routine. Other methods of solving the equations such as finite difference are also possible but in our experience the collocation method is far more efficient[13].

The use of the diffusion equation to represent the mass transfer rate increases the computing time requirements in comparison with the LDF model (eqn. 11) so in general the LDF approximation has been preferred.

Limitations of LDF Model

In order to examine the limitation of the LDF model, Nakao and Suzuki[12] solved the equations for a single adsorbent particle subjected to a periodic change in external sorbate partial pressure according to both LDF and diffusion models. By comparing these solutions they were able to show that the LDF model can provide a reasonable representation of the system behaviour provided the LDF coefficient (Ω in $k = \Omega D_c/R^2$) is adjusted as a function of cycle time, as shown in figure 2. At long cycle times Ω approaches the Glueckauf value of 15 but for shorter cycle times the value is substantially higher.

In order to examine the correspondence between the LDF and pore diffusion models in greater detail, Raghavan et al.[14] simulated a Skarstrom PSA cycle over a range of operating conditions using both LDF and diffusion models. The results, which are summarized in figure 2, are broadly similar to the results obtained by Nakao and Suzuki from their single particle analysis but there are significant differences. The cycle time is the most important variable determining the appropriate value for Ω. For long cycle times $\Omega \rightarrow 15$, in agreement with the results of Nakao and Suzuki and in accordance with the Glueckauf model. However, at short times, we find that Ω approaches a constant value in the range 30-40, rather than continuing to increase as cycle time is reduced.

As a simplified representation of the diffusion equation some authors have used the assumption of a parabolic concentration profile within the particle.[15,16] This approximation provides a good representation of a diffusion process at intermediate and longer times but it breaks down in the initial region. In fact the

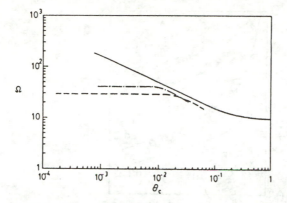

<u>Figure 2</u> Variation of Ω with cycle time. Nakao and Suzuki (single particle) ——; Raghavan et al. (PSA cycle) ---, --·--·.

parabolic profile approximation may be shown to be formally identical to the Glueckauf LDF model ($\Omega = 15$) so this approach does not have any real advantage over the LDF approach and is subject to the same limitations when the cycle time is short relative to the diffusion time (R^2/D_e).

<u>CO$_2$-He-Silica Gel</u>

The earliest attempt at a dynamic PSA simulation is due to Mitchell and Shendalman[17,18] who also studied experimentally the adsorption of traces of CO_2 from a He carrier onto silica gel in a Skarstrom cycle. Their model used the LDF approximation and is essentially as outlined in eqns. 1-9. Agreement between model and experiment was, however, poor. Re-examination of these results revealed that the same mass transfer rate constant had been assumed for both adsorption and desorption steps. Since the process is in fact controlled by intraparticle diffusion and molecular diffusion mechanism is dominant, one would expect $D_e \alpha 1/P$ so that the effective diffusivity during the adsorption steps should be decreased in proportion to the pressure ratio. With this modification the model provides a reasonable representation of the system dynamics[19].

<u>C$_2$H$_4$-He-4A/5A Zeolite</u>

In order to examine further the impact of mass transfer resistance Hassan et al.[20] studied the adsorption of a small concentration of ethylene from He in a PSA system with 4A or 5A zeolite adsorbents. Intracrystalline diffusion of C_2H_4 in 4A zeolite is relatively slow and is the rate controlling mass

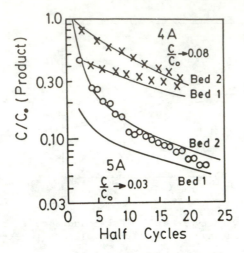

<u>Figure 3</u>   Comparison of theory and experiment for C_2H_4-He-4A/5A
showing approach to cyclic steady state. For 4A, $k \cong 10^{-3}$
s^{-1}, half cycle time = 450 s; for 5A; $k \cong 0.19\ s^{-1}$, half
cycle time = 100 s.

transfer step under the experimental conditions. By contrast, in
5A, intracrystalline diffusion is rapid so that pore diffusion
becomes the dominant rate process. To model the system we used
essentially the model outlined above with Langmuir equilibrium and
the LDF rate expression. The relevant kinetic and equilibrium
parameters were derived from experimental breakthrough curves for
adsorption and desorption measured under conditions similar to the
PSA operation. The LDF coefficients were corrected for pressure in
accordance with the relevant diffusion mechanism (macropore or
intracrystalline) and adjusted for cycle time according to figure 2.

The approach of the outlet concentration to the cyclic steady
state, for 4A and 5A zeolites, is shown in figure 3 from which it
is evident that the simplified dynamic model provides a reasonable
representation of the system behaviour. The effect of the large
difference in mass transfer resistance is clearly apparent. 5A
zeolite clearly gives better performance than 4A. The purity of
the 4A product may be increased by lengthening the cycle time but
only at the cost of reduced throughput.

3   AIR SEPARATION

The traditional PSA air separation process, which is widely
used for small scale production of oxygen, depends on the selective
(equilibrium) adsorption of nitrogen on a zeolite adsorbent

(generally 5A or 13X) – see figure 4.[21] If nitrogen is the desired product it is preferable to use an adsorbent which selectively adsorbs the oxygen. This can be accomplished with either a 4A zeolite or a suitable carbon molecular sieve adsorbent. These adsorbents have an effective micropore diameter close to the molecular diameter of O_2 and N_2 so that the slightly smaller O_2 molecule diffuses much faster than N_2 making an efficient kinetic separation possible. The equilibrium isotherms for oxygen and nitrogen on carbon sieve are almost identical, whereas on the 4A zeolite, as on 5A, N_2 is more strongly adsorbed (figure 4) so that kinetic and equilibrium effects are in opposite directions. In a kinetic separation the choice of operation conditions, in particular the cycle time, is critical since if the adsorbent bed is allowed to approach equilibrium the selectivity is lost.

Table 1: Summary of Kinetic and Equilibrium Parameters for Sorption of O_2 and N_2 on Bergbau CMS and Linde 4A Zeolite (300K).

Parameter	CMS	4A
K_{N_2}	9.0	8.2
K_{O_2}	9.0	3.8
$-(\Delta H)_{N_2}$ (kcal/mol)	3.8	4.3
$-(\Delta H)_{O_2}$ (kcal/mol)	3.8	3.2
$(D/R^2)_{N_2}$ (s^{-1})	1.17×10^{-4}	2×10^{-3}*
$(D/R^2)_{O_2}$ (s^{-1})	3.7×10^{-3}	0.3*
E_{N_2} (kcal/mol)	6.5	4.5
D_{O_2}/D_{N_2}	32	~ 150

* Diffusivity values extrapolated from low temperature measurements. Data of Ruthven and Hassan (CMS)[22] and Ruthven and Derrah (4A)[23].

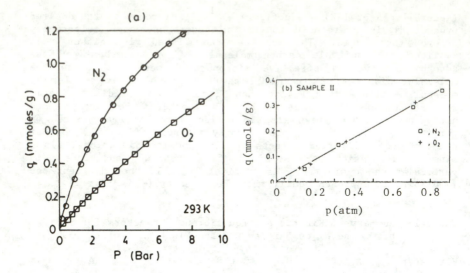

Figure 4   Equilibrium isotherms for O_2 and N_2 on (a) 5A Zeolite
and   (b) Bergbau CMS.

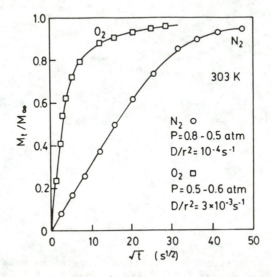

Figure 5   Uptake curves for O_2 and N_2 on Bergbau CMS at 303 K.

Kinetic and Equilibrium Parameters for O_2 and N_2 in CMS

Kinetic and equilibrium parameters for samples of carbon molecular sieve (Bergbau – Forschung) were determined by gravimetric and chromatographic methods.[22] The substantial difference in uptake rates with very little difference in equilibrium is illustrated in figure 5 while relevant parameters are summarized in Table 1, in which comparative data for 4A zeolite are also included.

Simulation of PSA Air Separation on CMS and 4A Zeolite

Although the traditional Skarstrom cycle can be used it is evident that the purge step is not really necessary in a kinetic separation. By proper choice of operating conditions and step times the bed may be made self-purging. The rapidly adsorbing species desorbs rapidly during blowdown while the slowly adsorbing species desorbs slowly and can provide sufficient desorbate to purge the bed. Various modified cycles have been suggested to take advantage of the self-purging nature of this type of system; one such cycle is shown schematically in figure 6. The operation of this cycle was simulated under typical operating conditions using the kinetic and equilibrium parameters for the molecular sieve carbon determined as outlined above.[13] The mathematical model was modified to allow for competitive adsorption and the variation in velocity through the bed as well as to account for non-linearity of the isotherm and competitive adsorption according to the binary Langmuir model. Figure 7 shows a comparison between the experimental and predicted performance in terms of product purity and recovery while figure 8 shows the effect of the diffusivity ratio as predicted from the model.

Results of a similar study of N_2 production over 4A zeolite have recently been presented by Shin and Knaebel[25,26]. A comparison between the performance of the CMS and 4A systems is shown in figure 9 in terms of the recovery vs product purity plot. Also included in this figure is a comparison between the performance of the CMS adsorbent operated on the Skarstrom cycle (fig. 1) and on the self-purging cycle (fig 6).[24,13] The advantage of the latter is clearly apparent since, over the entire range, for any given purity of the N_2 product, the recovery is significantly higher. Over most of the range the performance of the carbon sieve is superior to that of the 4A system but this situation is reversed at high product purities. In the high purity region the higher kinetic selectivity of the 4A sieve more than offsets the effect of the opposing equilibrium.

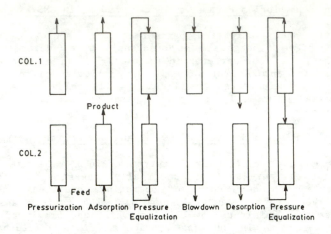

COL.1

Product

COL.2

Feed

Pressurization   Adsorption   Pressure   Blowdown   Desorption   Pressure
                             Equalization   Equalization

<u>Figure 6</u>   PSA cycle for N_2 production using carbon molecular sieve.

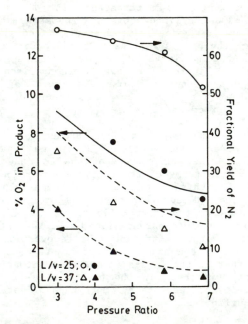

<u>Figure 7</u>   Performance of CMS N_2 production unit: comparison of
model predictions with experiment for two different
L/v ratios.

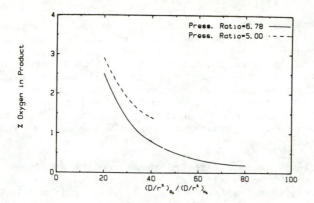

Figure 8   Effect of diffusivity ratio on performance.

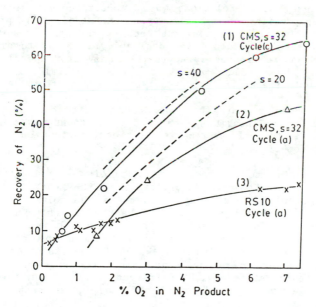

Figure 9   Comparison of performance of CMS and 4A zeolite (RS10)
processes for N_2 production.   Cycle (a) –Skarstrom,
cycle (c) – fig. 6. $s = D_{O_2}/D_{N_2}$.

4   HEAT EFFECTS IN PSA SYSTEMS

So far in our discussion of PSA systems we have assumed isothermal behaviour and neglected any effects arising from the heat of adsorption. In some systems, especially for bulk separations and in large columns which run almost adiabatically, the heat effects can be important. Under these conditions the modelling problem becomes more difficult since it is necessary to include the heat balance equations and to allow for the temperature dependence of the kinetic and equilibrium parameters. This problem was first addressed by Chihara and Suzuki[27] who studied the air-moisture-alumina PSA system.

More recently we have extended these studies and investigated the effect of different initial conditions[28]. When the heat balance is included the model equations become highly non-linear as a result of the exponential variation of the adsorption equilibrium constant with temperature. As a result we find that the model equations have more than one solution. Depending on the initial conditions (clean beds or beds equilibrated with feed gas) one may approach two different cyclic steady states, as illustrated in figure 10 for the air-H_2O-alumina system. The existence of the two different steady state solutions was demonstrated numerically by advancing the profiles arbitrarily beyond the steady state profile. The profiles were found to converge back to the same steady state from the opposite direction. The desirable steady state giving a clean high pressure product is approached only from the clean bed initial condition. Such behaviour has practical implication concerning both the control and start-up of PSA systems.

The existence of more than one steady state solution appears to result from non-linearity in the model equations and can occur even in an isothermal system if the equilibrium isotherm is sufficiently non-linear.

In attempting to demonstrate multiplicity experimentally we found that in a small scale laboratory unit the heat effects are severely attenuated by the heat capacity of the column, the heat conduction of the column walls and by heat loss, which is always difficult to eliminate in a small unit. However these effects can be allowed for in the model. Figure 11 shows the results of experiments carried out with C_2H_4-He-5A in a Skarstrom cycle. Although the temperature changes are small the existence of two different steady states with different product purities and different temperature profiles is confirmed by the experimental results.

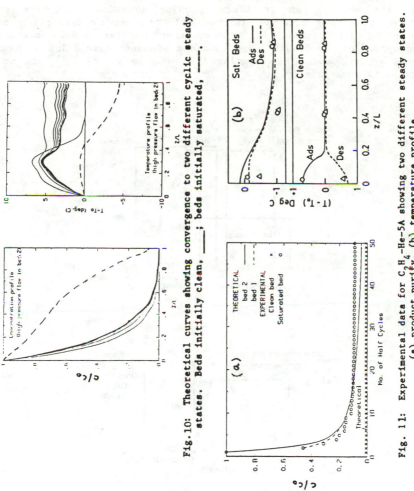

Fig.10: Theoretical curves showing convergence to two different cyclic steady states. Beds initially clean, ⎯⎯⎯; beds initially saturated, ⎯ ⎯ ⎯.

Fig. 11: Experimental data for C_2H_4–He–5A showing two different steady states. (a) product purity, (b) temperature profile.

5 CONCLUDING REMARKS

The models discussed here are among the simplest of the dynamic models since numerous simplifying approximations are used. Nevertheless the essential features of a PSA system are retained and the models can therefore provide useful qualitative or semi-quantitative guidance in the selection and optimization of operating conditions. It appears that similar models are in widespread use in industry and indeed a standard simulation package based on the model as formulated by Hassan et al.[13,24] is now available commercially.[29]

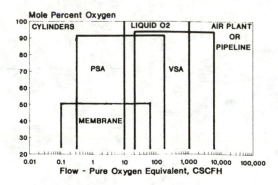

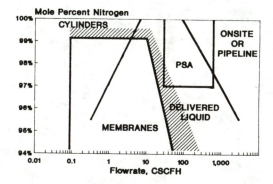

Figure 12   Comparative economics of air separation processes. (Spillman[30].)

In view of the focus of this conference on various methods of air separation it is pertinent to consider briefly how the economics of PSA compare with alternative technologies. Such a comparison, based on Monsanto data, is shown in figure 12. For O_2 production at moderate scales (< 1000 SCFH) PSA has a clear economic advantage while at somewhat larger scales the advantage shifts to liquid oxygen or a vacuum swing system (a variant of PSA). Membranes compete effectively only where a low purity oxygen product (< 50%) is required. For N_2 production where a relatively pure product is generally required, PSA is advantageous at somewhat larger scales (5000 - 100,000 SCFH) but at lower flowrates and lower product purities membrane systems offer superior economic performance. The crossover point at which PSA system outperforms a cryogenic distillation system has moved steadily to higher capacities over the last decade. However, it is still true that, for both O_2 and N_2 production at very large scales, the cryogenic process provides the most economic choice.

NOTATION

b	Langmuir equilibrium constant
c	gas phase concentration
c_o	concentration of sorbate in feed
D	diffusivity
D_e	effective diffusivity
D_L	axial dispersion coefficient
D_p	pore diffusivity
k	mass transfer coefficient in LDF model (eqn. 11)
K	Henry's Law equilibrium constant
P	total pressure
q	adsorbed phase concentration (local)
q*	equilibrium value of q
$\bar{q}$	value of q averaged over adsorbent particle
q_s	saturation limit
r	radial coordinate
R	particle radius
t	time
v	interstitial gas velocity
z	axial distance
ε	bed voidage
ε_p	particle porosity
Ω	constant in $k = \Omega D_e/R^2$.

REFERENCES

1. C.W. Skarstrom, U.S. Patent 2, 1958, 944, 627.
2. C.W. Skarstrom, p. 95, Vol. 2, "Recent Developments in Separation Science", N.N. Li ed., C.R.C. Press, Cleveland, 1972.

3.   D. Domine and G. de Montgareuil, French Patent 1, 1957, 223, 261.
4.   D. Domine, and L. Hay, in "Molecular Sieves", 1968, p. 204, Society for Chemical Industry, London.
5.   P., Wankat, Large Scale Adsorption and Chromatography, CRC Press, 1986, Boca Raton, Fla.
6.   R. Yang, "Gas Separation by Adsorption Processes", Butterworths, Stoneham, Mass, 1987.
7.   Y.N.I. Chan, F.B. Hill, Y.W. Wong, Chem. Eng. Sci., 1981, 36, 243.
8.   K.S. Knaebel and F.B. Hill, Chem. Eng. Sci., 1985, 40, 2351.
9.   G. Flores-Fernandez and C.N. Kenney, Chem. Eng. Sci., 1983, 38, 827.
10. E. Glueckauf and J.E. Coates, J. Chem. Soc., 1947, 1315.
11. E. Glueckauf, Trans. Faraday Soc. 51, 1965, 1540.
12. S. Nakao and M. Suzuki, J. Chem. Eng. Japan, 1983, 16, 114.
13. M.M. Hassan, N.S. Raghavan, D.M. Ruthven, Chem. Eng. Sci., 1987, 42, 2037.
14. N.S. Raghavan, M.M. Hassan, D.M. Ruthven, Chem. Eng. Sci., 1986, 41, 2787.
15. C.H. Liaw, J.S.P. Wang, R.A. Greenkorn, R.C. Chao, AIChE Jl. 1979, 25, 376.
16. R.G. Rice, Chem. Eng. Sci. 29, 1984, 1828.
17. J.E. Mitchell and L.H. Shendalman, Chem. Eng. Sci., 1972, 27, 1449.
18. J.E. Mitchell and L.H. Shendalman, AIChE Symp. Series, 1973, 69 (134), 25.
19. N.S. Raghavan, M.M. Hassan, D.M. Ruthven, AIChE Jl., 1985, 31, 385.
20. M.M. Hassan, N.S. Raghavan, D.M. Ruthven, AIChE Jl., 1985, 31, 2008.
21. G.A. Sorial, W.H. Granville, W.O. Daly, Chem. Eng. Sci., 1983, 38, 1517.
22. D.M. Ruthven, N.S. Raghavan, M.M. Hassan, Chem. Eng. Sci., 1986, 41, 1325.
23. D.M. Ruthven and R.I. Derrah, J. Chem. Soc. Faraday Trans. I 1975, 71, 2031.
24. M.M. Hassan, N.S.Raghavan, D.M. Ruthven, Chem. Eng. Sci., 1986, 41, 1333.
25. H-S. Shin and K.S. Knaebel, AIChE Jl., 1987, 33, 654, 24.
26. H-S. Shin and K.S. Knaebel, AIChE Jl., 1988, 34, 1409.
27. K. Chihara and M. Suzuki, J. Chem. Eng. Japan, 1983, 16, 53.
28. S. Farooq, M.M. Hassan, D.M. Ruthven, Chem. Eng. Sci., 1988, 43, 1017.
29. Anon, Prosep News, Newsletter of Prosys Technology, Feb. 1989.
30. R.W. Spillman, Chem. Eng. Prog., 1989, 85 (1), 41.

The Bed Dynamics of Pressure Swing Adsorption

C. N. Kenney

DEPARTMENT OF CHEMICAL ENGINEERING, CAMBRIDGE UNIVERSITY,
PEMBROKE STREET, CAMBRIDGE CB2 3RA, UK

Several aspects of conventional PSA were reviewed by Professor Ruthven in his lecture. In particular, the starting point was the Skarstrom two bed cycle and the difference between equilibrium and mass transfer models was emphasised. Mass transfer and diffusion effects play a major role in explaining nitrogen separation by carbon molecular sieves. Implicit in the analysis was the assumption that the adsorbent particle sizes used are sufficiently large, typically 1-3 mm, so that at the pressurisation and depressurisation rates used, the beds are isobaric, that is all points along a bed undergo the same pressure variation with time.

Here two different and complementary problems in PSA will be examined, together with a discussion of some of the experimental data recently obtained in our laboratory.

1) The factors influencing the shapes of the adsorption fronts which move through the adsorbent beds during the various process steps in PSA are important and explain why more complex cycles involving interchange of gas between beds are used in practice, to obtain oxygen rich streams.

2) A more recently developed variant of PSA is Rapid Pressure Swing Adsorption RPSA in which very short cycle times (1-3s) are used in a single bed of small particles (<500µ). High oxygen purities are obtained (>70%) in a single bed. How does this operation differ from conventional PSA?

These are both problems for which significant insight into gas separation, particularly air into its components, can be obtained by examining equilibrium models in which it can be assumed that at all times and points, in a column the gaseous and adsorbed phases are in equilibrium.

1 ADSORPTION FRONTS IN PSA

Crucial to the understanding of PSA is the way in which gas streams behave when one stream of a given composition displaces another. In isobaric displacement of pure oxygen by air, and air by oxygen, two very different types of column breakthrough are obtained, the sharp 'shock front' and the gentler 'simple wave'. This behaviour is observed in the analysis of the pressurisation of an adsorbent bed when air in the bed undergoes pressurisation (Fig 1).

Characteristic lines - single gas

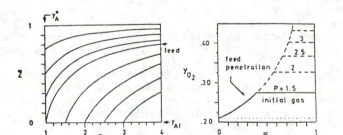

Characteric lines and concentration profiles during pressurisation.

Uniform displacement of initial gas : $y_A^o = y_{Af}^o = 0.21$

Fig 1

It will be seen that increasing bed pressure by feeding in air leads to an increasing oxygen rich plateau developing at the closed end of the bed and the length of this plateau decreases as the total bed pressure rises. This is confirmed by experiment (Fig 2). The concentration distribution at sample points in the bed followed by mass spectrometry in a single bed, cycled between pressurisation, product release and pressurisation, also agree adequately with the predictions of the model (Figs 3 and 4).[1]

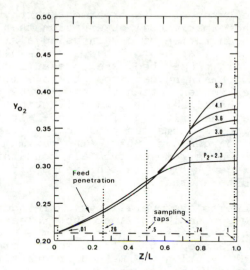

Experimental concentration profiles for the air pressurisation runs.

Fig 2

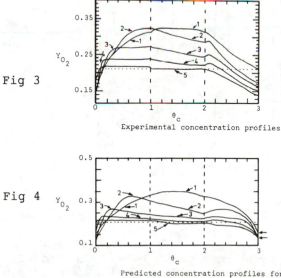

Fig 3

Experimental concentration profiles

Fig 4

Predicted concentration profiles for pressurisation, product release and depressurisation.

Numbers 1-5 indicate sample points

There are two features of this treatment worthy of
comment: (1) the experimental observation that the
maximum exit concentration of oxygen using a single bed
with maximum pressurisation to 5 bar, is 35-40% and (2)
the enrichment in a closed bed, when pressurised with air,
or other oxygen containing gas 'bootstraps', i.e. the
higher the oxygen concentration present in the bed before
pressurisation, the higher the oxygen fraction in the
oxygen plateau at the exit end of the bed.   It is well
known that two bed systems can produce a continuous
oxygen supply of oxygen over 90%. (The maximum achievable
if all the nitrogen is removed from air containing 1%
oxygen is 95% oxygen). This is achieved by transferring
gas from one bed to another.

When further steps to the overall cycle are added to the
three of (1) pressurisation, (2) product release and (3)
depressurisation, by transferring gas from a second bed,
the effects can again easily be followed in terms of an
equilibrium model by using the method of characteristics.[2]
In the purge step oxygen rich gas from bed 2 is used to
displace residual nitrogen from bed 1 (Fig 5).

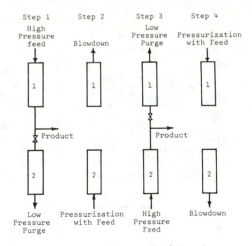

Fig 5.   A Two Bed Purge Cycle

If the entrance to bed 1 is then closed, the 'backfill',
that is partial repressurisation of bed 1 with oxygen rich
gas from bed 2 ensures that when bed 1 is pressurised with
air, the product that can be taken from bed 1 contains

much more oxygen than in the single bed system. Note the
existence of an optimum amount of gas to be transferred in
the purge step. Too little gas means residual nitrogen is
not completely expelled from the inlet of bed 1; too much
gas results in oxygen rich gas from bed 2 traversing the
length of bed 1, leaving the inlet of 1 and hence being
lost and wasted. These steps can be combined to ensure
there is continuous exit production of oxygen rich gas
(Fig 6).

BED | SIMPLE CYCLE

BED								
1	Product release			Depressurisation	Null	Pressurisation		
2	Depressurisation	Null	Pressurisation	Product release				

PURGE CYCLE

BED									
1	Product release			Depressurisation	Purge	Null	Pressurisation		
2	Depressurisation	Purge	Null	Pressurisation	Product release				

BACKFILL CYCLE

BED									
1	Product release			Depressurisation	Null	Backfill	Pressurisation		
2	Depressurisation	Null	Backfill	Pressurisation	Product release				

PURGE-BACKFILL (PB) CYCLE

BED										
1	Product release			Depressurisation	Purge	Null	Backfill	Pressurisation		
2	Depressurisation	Purge	Null	Backfill	Pressurisation	Product release				

Fig 6.   Some Two Bed Cycles

More refined analyses are possible by relaxing the
equilibrium model and allowing for mass transfer to and
within the zeolite pellets by mass transfer. The values
of the coefficients for oxygen and nitrogen $k_O = 0.3$ cm/s
and $k_{N_2} = 0.15$ cm/s were obtained by fitting the
experimental isobaric breakthrough curves using a 21%
O_2 - 79% N_2 mixture, pure nitrogen and pure oxygen
(Fig 7).

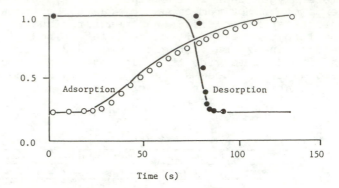

Fig 7.   Isobaric Breakthrough Curves in a Single Bed Used
          to Estimate Mass Transfer Coefficients

The relevant equations which are essentially of the
'linear driving force' form are discussed elsewhere and
can be used to examine the role of the backfill step,
which can replace many of the functions of the
pressurisation step in that the high oxygen concentration
front is formed at the centre of the bed during the
backfill step.   The pressurisation step then only needs to
push the front towards the product end.   The model shows
good agreement with experimental results, the best being
around 0 to 200 kPag backfill pressure where the sharp
rise in oxygen concentration occurs, and is followed by a
concentration 'plateauing'.   Above 200 kPag backfill
pressure, the theory predicts a small decrease in the
product oxygen concentration with increasing backfill
pressure.

Concentration Profiles in the Bed [3]

Further theoretical analyses can be used to show the bulk
gas phase oxygen concentration profiles in the bed at the
end of the pressurisation, backfill and depressurisation
steps at steady state for a number of different backfill
pressures, as shown in Fig 8.

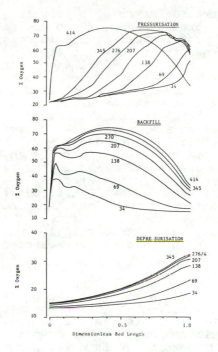

Fig 8.   Simulated Oxygen Profiles for the
Parts of the Cycle in a 2-Bed Process
for Different Back-fill Pressures

The maximum bed pressure in these simulations is 410 kPag.
A modified linear driving force model is employed with the
equations being solved by orthogonal collocation.

(a) Pressurisation

In the 35 kPag backfill pressure case, the amount of feed
gas in the pressurisation step is much larger than that in
the backfill step. As the feed gas is at a fixed oxygen
concentration which is lower than the maximum in the bed,
the oxygen front in the bed is pushed towards the product
end (Z = 1), forming a high oxygen concentration front.
The large amount of feed gas causes the oxygen
concentration front to be pushed right to the product end
and merges with the maximum left behind after the backfill
step to create a maximum at Z = 1. For the 70 to 410 kPag
backfill pressure cases, the ratio of the amount of feed
gas in the pressurisation step increases the value of the

oxygen concentration maximum, but it is unable to create a maximum at the product end. Moreover, for the 210 to 390 kPag backfill pressure cases, the incoming oxygen feed creates an oxygen front from the steep profile near the $Z = 0$ end, left by the backfill step, but is unable to push it far enough so that it merges with the backfill maximum. The new oxygen front creates a second maximum with an oxygen concentration higher than that of the first maximum. For the 35 to 100 kPag backfill pressure cases, the pressurisation step pushes the oxygen concentration maximum so that it lies between $Z = 0.7$ to 1. For the 410 kPag backfill pressure case, there is no separate pressurisation step as 410 kPag is the final bed pressure and the maximum oxygen concentration lies at $Z = 0.4$. Therefore, over the 390 to 410 kPag backfill pressure range, the position of the highest maximum oxygen concentration moves from $Z = 0.7$ to 0.4 and this accounts for the slightly larger theoretical drop in oxygen concentration in the product, as measured experimentally.

(b) <u>Backfill</u>

The backfill step is similar to the pressurisation step except that it is effected in the opposite direction. The feed for the backfill step sees a decreasing gas phase oxygen concentration as it moves down the bed and a shock wave is formed whose front is diffused by the effects of axial dispersion and mass transfer resistance. Nevertheless, it can be seen from the profiles that the shock wave fronts are still steep near the region of $Z = 0$ to $Z = 0.05$. The shock wave fronts increase in height rapidly for backfill pressures between 30 to 100 kPag backfill pressure, but change slightly for higher backfill pressures of 120 to 410 kPag.

For 30 to 100 kPag backfill pressure cases, there are two maxima in the oxygen concentration profile. The first maximum near $Z = 0$ arises from the shock wave caused by the backfill feed gas pushing the oxygen in the bed towards the $Z = 0$ end. The second maximum around $Z = 0.2$ is brought about by the change in oxygen concentration of the backfill feed. When product gas leaves a two bed rig, the product release step is to split into two stages. In the first stage, product is released solely into the product line, while in the second stage, the product is released into both the product line as well as the other bed, as feed for the backfill step. The product of the second bed, used as the backfill feed thus loses its high oxygen concentration portion first to the product. Therefore, after a short period, the maximum oxygen concentration in the feed bed is pushed out of the bed and the product stream has an oxygen concentration that is lower than that in the bed it is backfilling. The feed into this bed (the feed is decreasing in oxygen concentration with time) thus creates a weak second shock front with a diffuse profile being formed away from the second maximum towards the $Z = 1$ end. For the 200 to 410 kPag backfill pressure cases, the feed into the bed is at a high enough concentration initially to create a diffusing front which merges into the first maximum to form an oxygen concentration profile with only one maximum. The position of the maximum moves from around $Z = 0.38$ to $Z = 0.42$ over the 200 to 410 kPag backfill pressure range. The largest change in oxygen concentration over the whole bed occurs over the 30 to 180 kPag backfill pressure range, the change being much less for the 180 to 410 kPag backfill pressure range.

(c) <u>Depressurisation</u>

The oxygen profile at the termination of the depressurisation step becomes progressively steeper as the backfill pressure preceding it, is increased. The oxygen profile changes rapidly between 30 to 130 kPag backfill pressure size, but changes slowly in the 180 to 410 kPag backfill range. This similarly coincides with the very steep change in product oxygen concentration with backfill pressure as shown in Fig 8. The profile gets progressively much steeper towards the product end of the bed ($Z = 1$), showing the purging effect of the depressurising flow as the higher oxygen concentration purges the nitrogen out of the bed. The small high oxygen front left at the product end ($Z = 1$) of the bed, after the product release step, acts as a purge stream. Without the backfill step, the pressurisation step alone is unable to create an oxygen front large enough so that a small portion would be left behind to purge the bed after the product release step. However, even this small portion is inadequate to fully purge the bed and the oxygen concentration at the ($Z = 0$) end is still much lower than the feed oxygen concentration. The use of a purge step at the end of the depressurisation step so that the oxygen concentration at the ($Z = 0$) end is close to the feed oxygen concentration would enhance both the product oxygen concentration and the oxygen recovery.

2. RAPID PRESSURE SWING ADSORPTION

In recent years there has been a developing interest in a variant of PSA in which the adsorbent particles are much

smaller than those used in traditional PSA, that is with
diameters in the range 100μ - 500μ (Fig 9).

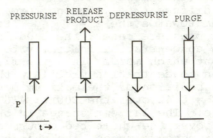

Figure 1.1 – A Conventional Pressure Swing Adsorption Cycle

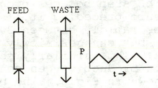

Figure 2.1 – The Rapid Pressure Swing Adsorption Cycle

Fig 9.   Comparison of Pressure-Time Variation
          in PSA and RPSA

The result of this is that in the pressurisation and
depressurisation steps it cannot be assumed that the bed
is isobaric, and pressure changes which are functions of
both time and distance must be allowed for. Incidentally,
these transients move relatively slowly and are similar to
the pressure variation in transient gas flow in a
capillary and do not travel at sonic velocities. Greater
outputs of oxygen enriched gas can be obtained in unit
time with RPSA compared with conventional PSA, although
there is an energy penalty.

By alternating the pressurising and depressurising steps
in a single bed of zeolite particles with a cycle time of
1 - 3 seconds, it is possible to obtain a continuous
stream of exit gas enriched in oxygen (> 70% O_2) from air.
The small particles result in the pressure at the exit end
of the column always lagging behind the instantaneous
inlet pressure. Enrichment of the gas at the exit end of

the column in the pressurisation step occurs largely as in isobaric PSA, but in the depressurisation step, near the exit end of the column, product gas is leaving the end of the column, and oxygen enriched gas is flowing back towards the bed inlet, displacing and purging nitrogen rich gas which has accumulated in the column near the column inlet. The separation mechanism differs from that in PSA in the flow-pressure variation which leads to the enrichment of a pocket of gas as it moves through the column with a net forward (backwards and forwards) movement in each cycle. The analogy with a ratchet helps to visualise the process and it should be noted that velocity of travel of such a pocket is significantly less than superficial velocity of gas through the bed (Fig 10).

R.P.S.A - How does it work ?

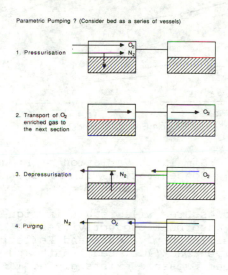

Parametric Pumping ? (Consider bed as a series of vessels)

1. Pressurisation
2. Transport of O_2 enriched gas to the next section
3. Depressurisation
4. Purging

Note. "Ratchet effect" : Time for 'slug' of gas to traverse bed longer than pressurisation-depressurisation cycle.

Fig 10

When modelling RPSA we have to combine the usual equations describing the total and component mass balances with an equation relating flow to pressure drop. Darcy's Law provides a good start.

$$v = -\frac{k}{\mu}\frac{dp}{dz}$$

Solving these equations, even assuming that we can ignore
gas-solid mass transfer, gives results which demonstrate
the major features of RPSA, in particular, exit oxygen
concentrations of 60 - 70%. Other features to come from
the simulations are the prediction of an optimum bed
permeability, k. Major process variables which include
the feed pressure, exhaust pressure and product flowrate,
all of which affect either or both the product purity and
recovery. As might be expected, oxygen concentration
falls as the amount of product is increased (Fig 11).[4]

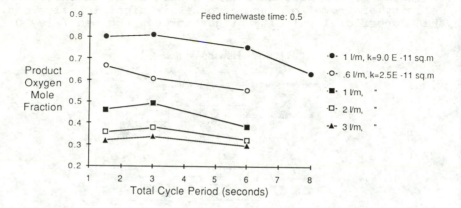

Fig 11.   Simulated Variation of Product Oxygen
           Concentration with Product Rate for two bed
           Permeabilities

Higher feed pressures will result in higher product
purities and adsorbent productivities at the expense of
compressive efficiency. Jones and Keller (1981) state
that for a given bed (i.e. bed length and particle size) a
particular feed pressure exists giving maximum
productivity at the highest recovery. The feed pressure
will affect equilibrium adsorption quantities and the
characteristics of the developed cyclic pressure
gradients. Such gradients, in turn, will affect internal
purging rates (counter-current) and equilibrium desorption
quantities, and thus the regenerative properties of the
process. Furthermore, an increase in the feed pressure
will result in an increase in feed and exhaust flowrates,
and therefore different feed pressures will have different
optimum pressure cycles. For a given feed pressure and
pressure cycle, there is an optimum product flowrate which
maximises both the recovery and compressive efficiency.[5]

Turnock and Kadlec (1971) have shown how the product purity is a decreasing function of the product to feed rate ratio (note: the feed rate is a dependent variable based upon feed pressure, pressure cycle and the product flowrate). The product flowrate will again affect the characterisitics of the pressure gradients formed within the bed.

A decrease in exhaust pressure will result in an increase in product recovery, as demonstrated by Keller and Jones in a process separating hydrogen from hydrogen-methane mixtures. Low exhaust pressures will result in sharper and greater pressure drops within the bed, and therefore greater releases of the more adsorbable component. This in turn will reduce the partial pressure of the less adsorbable component, since the total exhaust pressure is fixed, thus increasing the product (i.e. less adsorbable component) recovery. Hence it is common to operate RPSA processes at atmospheric exhaust pressure, although pressures below atmospheric may, theoretically, further increase recoveries, albeit at the expense of greater energy demands. Particle size, as investigated by Pritchard and Simpson (1986) and Dankworth (1987), is shown to have a large effect on both product recovery and purity. Particle size affects the degree of flow resistance and hence the transmission of the pressure waves within the bed.

It is clear that more research on RPSA is necessary to exploit to best effect the simplicity of the single column but sufficient is already known to indicate there is a niche of uses where it can be employed to advantage.

Conclusions:

PSA is an advanced separation process being the method of choice for an increasing number of applications. It demonstrates the effective interaction between adsorbent development, high quality engineering, and the computer simulation of a complex non-steady state process. It has rapidly progressed to being a mature technology but one in which theoretical and practical advances are still possible. The work described in this lecture owes much to the efforts of my research students over the last decade, F. Florez-Fernandez, N.F. Kirkby, J-L. Liow, D. Dankworth and E. Alpay, and their insights and contributions are gratefully acknowledged.

286   *Separation of Gases*

References

1) G. Flores-Fernandez and C.N. Kenney. The modelling of
   the pressure swing air separation process.
   Chem. Eng. Sci., 1983, <u>38</u>, 827.

2) N.F. Kirkby and C.N. Kenney. The role of process
   steps in PSA cycles. "Fundamentals of Adsorption".
   Engineering Foundation Conference, California, 1986.

3) J.-L. Liow. Air Separation by Pressure Swing
   Adsorption. PhD Dissertation, Cambridge, 1986.

4) D.C. Dankworth. RPSA for Air Separation. CPGS
   Dissertation, Cambridge, 1987.

5) E. Alpay. Air Separation by RPSA. CPGS
   Dissertation, Cambridge, 1989.

6) R.J. Jones and G.F. Keller. Pressure-Swing
   Parametric Pumping. A New Adsorption Process.
   J. Separ. Proc. Technol., 1981, <u>2</u>, 17.

7) P.H. Turnock and R.H. Kadlec. The separation of
   nitrogen and methane via periodic adsorption.
   AIChEJ, 1971, <u>17</u>, 355.

8) C.L. Pritchard and G.K. Simpson. Design of an oxygen
   concentrator using the RPSA principle.
   Chem. Eng. Res. Des., 1986, <u>64</u>, 5.

Isotope Separation in the Nuclear Power Industry

C. Whitehead

BRITISH NUCLEAR FUELS PLC, CAPENHURST WORKS, CHESTER, CHESHIRE, UK

1.   INTRODUCTION

This contribution to the Conference addresses the requirements and techniques of Isotope Separation. An isotope of an element has the same chemical properties as other isotopes of the element, having almost identically the same electron configuration, but exhibits a different Atomic Weight which reflects an excess or deficiency of neutrons in the nucleus. This definition is almost true but not quite. Low Atomic Weight isotopes do exhibit noticeably different chemical properties which can be used to enrich one isotope with respect to others.

There is a market demand for many isotopes at the gram to kilo-gram level for uses in nuclear medicine and in industry for tracing and labelling techniques and in the manufacture of radiation sources for on-site radiography. The range of techniques to produce those small-quantity isotopes is very broad.

The nuclear industry has the largest demand for enriched materials. Heavy water, D_2O, is required for Canadian CANDU reactors at a rate of many tonnes per year, but the largest demand, measured in thousands of tonnes per year, is for uranium enriched in the U235 isotope for the manufacture of fuel used by Civil Nuclear Power Programmes in Advanced Gas Cooled Reactors (AGR), Boiling Water Reactors (BWR) and Pressurised Water Reactors (PWR). This paper will describe and compare methods of uranium enrichment for this purpose. The processes described will be the existing Gaseous Diffusion and Gas Centrifuge methods together with Laser Isotope Separation techniques which may offer future economic potential.

2.   ISOTOPE ENRICHMENT IN THE CIVIL NUCLEAR POWER
       INDUSTRY

The majority of reactors in the world require fuel in
which the uranium content has a U235 concentration greater
than that found in naturally occurring ore. The dominant
isotope in naturally occurring uranium is U238 with a
concentration about 99.3% with the U235 isotope at about
0.7% making up the remainder. It is the U235 isotope that
is thermally fissile and the fission of U235 produces most
of the power in a nuclear reactor. Reactor design
benefits from the use of fuel in which U235 has an
isotopic concentration in excess of 0.7%, and the majority
of reactors in the world require fuel with U235 isotopic
concentrations between about 2 and 4.5% depending on
reactor type and on the position of a fuel rod within the
reactor core.

The Western World demand for fuel with this enhanced
isotopic concentration is in excess of 5000 te per year
with an enrichment value of about £1.5B/yr and which
yields a power output of about 250 GWe.

The total Western World installed enrichment capacity
is about 7000 te of fuel/yr and is dominated (~85%) by
Gaseous Diffusion plants in the USA and in France and the
remainder dominated by Gas Centrifuge plants in Europe.
The excess total capacity relative to demand results in
both the American and French plants being operated below
full capacity.

Having identified the degree of enrichment required
and set the scale of production, the remainder of the
paper will describe the physics and engineering aspects of
Gaseous Diffusion and Gas Centrifuge plants, which today
produce the enrichment required for the World Civil
Nuclear Power plants, and a description of Laser Isotope
Separation, a potential future competitive process, will
also be given.#

REFERENCES

#Extensive bibliographies are given in:

1.   S Villani, Uranium Enrichment, Springer-Verlag, 1979
2.   S Whitley, Review of the Gas Centrifuge until 1962
               Reviews of Modern Physics, Vol 56, 41, 1984
3.   S Whitley, Isotope Enrichment by Diffusion and by the
               Centrifuge, Nuclear Energy, 27, 349, 1988

3.   GASEOUS DIFFUSION

The molecular weight dependence on the rate of diffusion of a gas through a porous barrier was demonstrated by Graham in 1846 and was used in 1920 by Aston to enrich the Ne22 isotope during his work on the isotopic constitution of the elements.

For a binary mixture of molecules with identical chemical properties but of different isotopic composition an elementary separation factor, α, can be defined as the ratio of the relative isotopic concentration after diffusion through a barrier compared to the ratio prior to diffusion.

$$\text{Then } \alpha = \sqrt{\frac{M_H}{M_L}} \tag{1}$$

where M_L and M_H are the molecular weights of the lighter weights of the lighter and heavier molecules, respectively.

In the case of uranium hexafluoride, UF_6, the gas invariably used in this process, $\alpha = 1.00429$. This is the theoretical maximum value and in practice is reduced by collisions between molecules and the pore walls, by pore size and geometry and by back pressure. These effects reduce α to an achievable value of about 1.0022.

The nature of the porous barriers and their method of construction are still closely guarded commercially. Diffusion theory demands that the pore diameter should be less than the mean free path between collisions of UF_6 molecules at pressures of about one atmosphere, ie less than about 20 nm and with lengths in excess of 20,000 nm. Such thin barrier material then requires mechanical support from a substrate. The corrosive nature of UF_6 vapour severely limits the choice of materials for barriers and substrates; some have been developed using aluminium, alumina, nickel and gold- and zinc- silver alloys.

The value of $\alpha = 1.0022$ implies that passage of uranium hexafluoride through a membrane could increase the isotopic concentration of U235 only from 0.7% to 0.7015%; it is evident that to achieve a product concentration of some few percent it will be necessary to repeat this process a large number of times in a cascade. Fig 1 shows part of a cascade with the flows indicated.

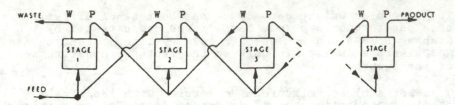

FIGURE 1   The principle of a diffusion cascade with
                      recyling between stages.

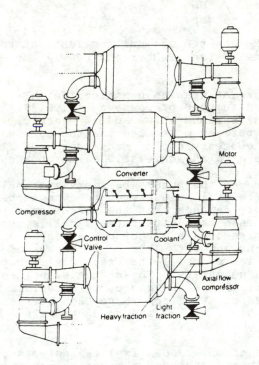

FIGURE 2   A practical realisation of diffusion
                      plant staging.

In an ideal cascade there is no remixing of materials with different isotopic concentrations at the points of confluence and the number of separation units in each stage is optimal.

The shape (ie the number of separation units per stage) and size of a cascade depends on two parameters: the final product flow required and the isotopic concentration required for that final product. Additionally, of course, the cascade should be close to ideal for reasons of economy; this latter requirement is usually achieved by "squaring-off" an ideal cascade to achieve an acceptable approximation to ideality.

The number of stages required to achieve a certain isotopic abundance enrichment is given by:

$$s = \frac{\ln \left(\dfrac{\text{Product concentration}}{\text{Feed concentration}}\right)}{\ln (\text{simple stage enrichment factor})} \qquad (2)$$

As an example, for a feed concentration of 0.7%, a desired product concentration of 3.5% and a simple stage enrichment factor of 1.0022 then 732 enrichment stages would be required.

Large gaseous diffusion plants have been built in the USA and in France. The French Plant at Pierrelatte has a capacity to produce enriched UF_6 sufficient for more than 2000 te of reactor fuel per year. The largest separation units have 2.5 MWe compressor units which handle 600 te of UF_6 per hour and contain more than 2000 m^2 of membrane.

The whole plant has 1400 stages and, with compressors and coolers between each stage, has a power demand of 3000 MWe which is supplied by four on-site Pressurised Water Reactors. The power demand for this enrichment process accounts for more than 75% of the enrichment costs.

Such a large plant also has a large "hold-up" of about 1000 te UF_6; this is the quantity of UF_6 in the plant at equilibrium. This gas cannot be introduced instantaneously and equilibrium must be achieved before economic take-off of product can occur; the equilibrium time for such a large plant is one to two months.

The high energy cost and lack of flexibility in achieving a given product are two of the disadvantages of the gaseous diffusion process.

This brief introduction to the gaseous diffusion method of enriching uranium has not mentioned many of the severe difficulties posed by the corrosive nature of UF_6 and the necessary selection of materials for the miles of pipework, for the compressors and coolers, the required leak tightness in view of the toxicity of UF_6 and of corrosive HF gas rapidly produced by the reaction of UF_6 with water or water vapour and the complex design of cascades to achieve the highest possible economic efficiency.

In spite of these problems the process produced 100% of the worlds enriched uranium for Civil Nuclear Power Programmes until the mid 1970's and still satisfies about 85% of the demand.

4.   THE GAS CENTRIFUGE ENRICHMENT PROCESS

Separation of mixtures of gases with different molecular weight using gravitational or centrifugal fields can be dated back to the latter years of the last century and after the demonstration of the existence of elemental isotopes Lindemann and Aston suggested that isotopes could be separated or enriched using these methods. Success only came in 1936 after development of the vacuum centrifuge in which rapidly rotating cylinders are suspended in a vacuum enclosure with devices to introduce and withdraw gases.

The principle of the method can be seen from the equation relating the pressure of a gas in "solid body rotation" within a rotating cylinder as a function of radius from the axis:

$$P(r) = P(o) \exp (Mr^2w^2/2RT) \qquad\qquad (3)$$

The influence of molecular weight is immediately clear and for a binary mixture the simple equilibrium enrichment factor α can be defined:

$$\alpha = \underline{\text{Abundance of heavy isotope at periphery}}$$
$$\text{Abundance of heavy isotope at axis}$$

$$= \exp \frac{(M_H - M_L) (wr)^2}{2RT} \tag{4}$$

It is seen that the separation is now dependent on the mass **difference** for this process and not on the mass **ratio** as for many other methods (q.v. gaseous diffusion). Such a method then has great advantage for the separation of heavy isotopes. For a mass difference of 3 units and a peripheral speed of 450m/s, $\alpha = 1.13$ which compares most favourably with $\alpha_{max} = 1.0043$ in the case of Gaseous Diffusion discussed earlier.

But all this potential cannot be used as at such speeds the interior of the centrifuge (from whence one would like to withdraw the material (UF$_6$) enriched in the desired U235 isotope) is at a very low pressure, less than 10^{-5}mm Hg on axis with a wall pressure approaching 100 mm Hg.

However the full potential can be retrieved by setting up an axial counter-current flow which then allows product and tails gas to be extracted from opposite ends of the centrifuge and, with correct design, at pressures suitable for injection into the following stage without an energy demand for re-compression and cooling. A temperature gradient along the length of the centrifuge will also generate an axial counter-current; in simple terms the gas at the hotter end will "rise", ie move towards the axis, and thus stimulate a counter-current flow. In practical centrifuges the counter-current is generated by a combination of scoop drag and temperature gradient effects and must be streamlined so that no turbulent mixing occurs. To achieve these conditions in an optimally controlled manner can be difficult. Figure 3 shows a schematic centrifuge.

The enrichment potential of the centrifuge has been exploited both by increasing the length and increasing the peripheral speed, but restrictions are found which arise from limitations in material properties. Improvements through the exploitation of the velocity-squared term find a limitation in that for a thin rotating cylinder the maximum value of velocity-squared is proportional to the ultimate tensile strength of the cylinder material; exploitation of the length term finds limitation arising in the dynamic response of long rotating cylinders and the

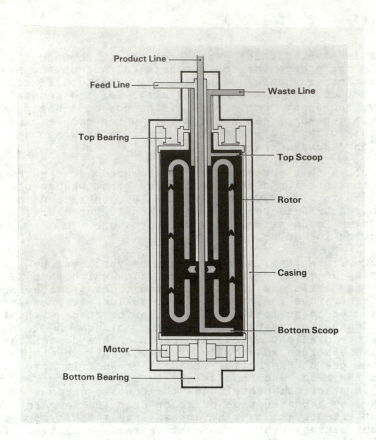

FIGURE 3   Schematic drawing of a centrifuge showing
                    injection and extraction of working gas
                    and the axial countercurrent.

existence of critical frequency phenomena which generate
requirements for high axial elastic modulus of the
cylinder material and requires bearing/damper systems with
high stiffness, freedom from cavitation and exhibiting low
drag and long life.

Within the URENCO/CENTEC International collaboration
between the United Kingdom, Holland and West Germany, the
development of centrifuges has indeed followed the
advances in materials. Recent materials such as
fibreglass, Kevlar and carbon fibre (all developed for the
aerospace industry) are capable of running at speeds in
excess 900 m/s, but without any gas (UF$_6$) loading or
protection against corrosion. Bearing development and the
development of the molecular pump to reduce the pressure
in the gap between rotor and cylinder have reduced drag
losses such that, at 1000 rev/sec, drive powers of only
some watts are required for each centrifuge and failure
rates of centrifuges have been reduced to only a few
tenths of one per cent per year.

The Urenco plants contain many tens of thousands of
centrifuges with an enrichment output sufficient to fuel
about twenty-four 1000 MWe PWR reactors or equivalent
every year. Fig 4 shows a view of an early enrichment
plant built at Capenhurst in the United Kingdom; the
centrifuges are grouped together in blocks which are in
turn connected in cascade.

In addition to the gas dynamics studies described
earlier, the development of centrifuges also involves
considerable investigation into stress analysis of the
many rotating components, into creep and stress-rupture
tests and extensive experimental verification of design
before the mass production of centrifuges for plant is
undertaken.

Within the Urenco/Centec collaboration the lapsed
time between first spinning of a single machine and the
first operation in plant is about 7 years for each new
generation of centrifuges.

5.   LASER ISOTOPE ENRICHMENT

This technique is characterised by the use of the
interaction of light (photons) with material (atoms or
molecules) and is an extension of the technique of
Photo-chemistry which has been studied since the last

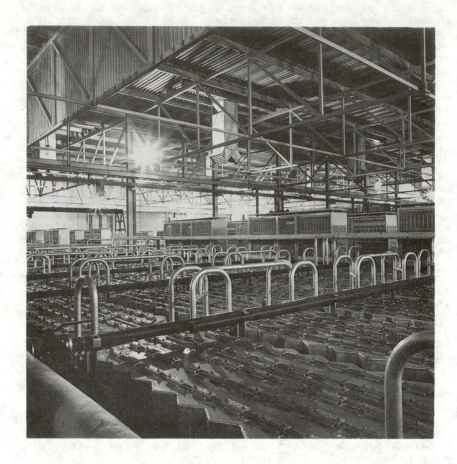

FIGURE 4   A view of centrifuge plant at BNFL
                         Capenhurst.

century. The intention and rapid development of lasers
resulted in many publications and patents in the early
1970s and subsequently. The real potential offered to
isotope separation was firstly the very narrow bandwidth
of laser light and its implied promise and secondly the
prospect of "cheap" photons which could result in an
economically-attractive process.

A number of isotopic enrichment processes have been
proposed: for example, the Atomic process operating with
uranium vapour, the Molecular route using uranium
hexafluoride and the CRISLA process again using uranium
hexafluoride, but with a selective process different to
that of the Molecular route. For the purpose of this
presentation attention will be directed only to the Atomic
process.

The first consideration is to understand why the
interaction with materials should allow selection between
isotopes. This selectively arises from the dependence of
the energy of excited atomic levels on the mass of the
nucleus. In the case of the uranium atom the precise
energies of most excited states of the atom differ by
about one part in one hundred thousand between the cases
where the nucleus has a mass of 238 or 235 atomic mass
units. Lasers exist which produce light beams in which
all the photons have energies equal within one part in ten
million. The capability thus exists to selectively excite
one or other of the isotopes if the laser output could be
tuned exactly (1 part in 10⁵) to induce a transition
between two isotope-specific energy levels.

Figure 5 shows this isotopic shift when a narrow band
laser is scanned across a particular atomic excitation
transition in the atoms in uranium vapour. The peak on
the left corresponds to the excitation in the U238 atoms
in the vapour and the multiplet on the right corresponds
to excitation of that same transition in the U235 atoms.
Two things should be noted: firstly, the excitation curve
is more complex for the U235 atom than the U238 atom; this
arises from the different spins of the U238 and U235
nuclei. J = O for U238 and J = 7/2 for U235, this
splitting phenomenon is Hyperfine Structure (HFS). The
second point to be noted is that the response level
between the two excitations falls almost to zero showing
that, as hoped, excellent discrimination between U238 and
U235 can be achieved in this excitation. This is not
sufficient though as both species would remain in their
excited states for only 10's to 100's of nanoseconds after

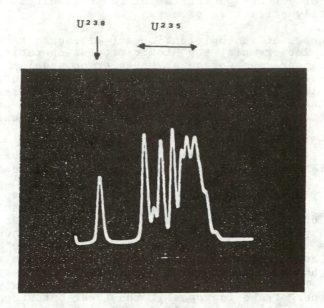

WAVELENGTH

FIGURE 5   A laser scan shows the isotope shift
                 between the U238 and U235 atoms in
                 uranium vapour and also shows the
                 Hyperfine structure of U235.

which time they would have decayed to their isotopically indistinguishable ground states.

This selective excitation process must be carried further, to achieve separation, by using other specifically tuned laser beams which can excite the atoms of interest to higher states (like climbing a ladder) and eventually to ionisation. If this is achieved then in the final state one would have unexcited U238 atoms plus ionised U235 atoms. The U235 ions can then be collected before they are neutralised by collisions or by capturing electrons.

Much spectroscopy has had to be carried out to fund "ladders" for which the "steps" have high transition strengths, exhibit adequate isotope shifts with minimal HFS and form states with life times long enough (> 100ns) to allow the "ladder" to be "climbed" with efficiency. The model of "ladder climbing" is far from appropriate to calculate the efficiency and selectivity of ionisation. A full Quantum mechanical model of the simultaneous interaction of the photon fields with the atom and its states must be developed before predictions can be made of the influence of transition strengths, polarisations, fluences and magnetic and electric fields on the efficiency and selectivity of the process.

As described above this process would allow perfect separation; however processes of collision during collection and geometric and efficiency difficulties deny this possibility.

The process can be described thus:

a.   produce unexcited uranium vapour

b.   selectively ionise the U235 atoms in the vapour

c.   collect the U235 atoms as efficiently as possible and minimise the collection of U238 atoms.

These principles of the process are indicated schematically in Fig 6 and a simplified illustration of a plant module in Fig 7.

Uranium is not the easiest metal to vaporise. Liquid uranium is close to the alchemists "alkahest", that

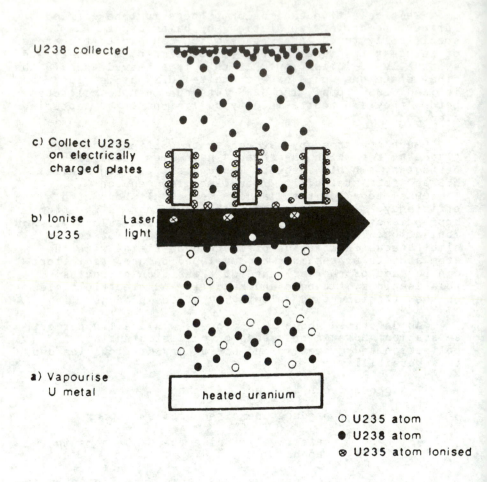

FIGURE 6   The principle of Atomic Laser Enrichment.

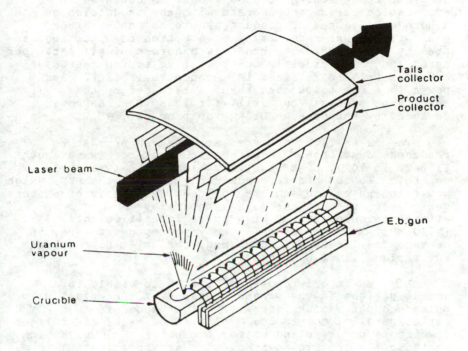

Tails collector

Product collector

Laser beam

Uranium vapour

Crucible

E.b.gun

FIGURE 7   Simplified illustration of the actual process.

is it tends to dissolve everything. Techniques do exist
in which the liquid uranium is confined in a thin shell of
solid uranium, a "skull", which in turn can be readily
contained in, for example, a cooled copper crucible.
Considerable power must be applied to evaporate the liquid
uranium; temperatures of 3000°C must be achieved if
evaporation rates commensurate with the production of some
hundreds of tonnes of product are to be achieved per
year. The power requirement of electron beam heating, at
present the preferred method, is hundreds of kilowatts per
meter length of crucible and some five to ten metres of
crucible will be required in a conceptual plant. Such
power is a few times more than is required by the
centrifuge process, but still far less than the power
demanded by the Gaseous Diffusion process.

The laser light to irradiate the vapour must be
produced to stringent requirements of precise wavelength
(1 part in 10^5), high stability (1 part in 10^6),
controllable and stable bandwidth (to cover the HFS), high
power and pulsed to achieve the required number of photons
per cm^2 ($> 10^{16}/cm^2$ per pulse) to achieve
efficient ionisation, a high pulse repetition rate
(> 20 kHz) to irradiate all the vapour produced and, of
course, the lasers (of which many tens will be required
for a plant) must be rugged.

At present only one laser system is viable for
investigations in the parameter range of conceptual plant:
pulsed copper vapour lasers (CVLs) arranged in
series/parallel to produce amplified beams of laser light
(a mixture of green and yellow lines for the CVLs) which
excite cells containing dyes (such as Rhodamine 6G) in
solution and these cells will then amplify light produced
at very low power levels, but precisely tuned to one of
the excitation steps, which is injected into them. A
chain of such dye amplifiers can amplify the power by a
factor $> 10^5$ to the levels required ($> 10^{16}$
photons/cm^2 in a 100 nsec pulse) without violating the
one part in 10^6 stability requirement.

This then leaves collection of the selectively
ionised atoms. The collection process is made difficult
by the fact that the ionisation process produces a plasma
of ions and electrons; this plasma, of density about
10^{11} ion pairs/cm^3, resists the collection of
the ions by the use of simple electrostatic fields due to
the Debye shielding phenomena. This resistance is not
absolute, but slows down the collection process so that

FIGURE 8   Part of the Copper Vapour/Dye Amplifier
                  Laser System at Harwell

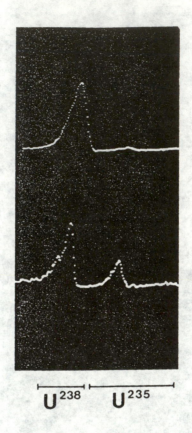

U^{238} U^{235}

FIGURE 9   Alpha particle spectra of natural uranium
                         (upper trace) and the spectrum from a
                         sample deposited from vapour irradicated
                         with lasers (lower trace). An
                         enhancement of a factor about nine is
                         achieved.

the efficiency of collection is severely reduced by recombination or by momentum or charge exchange scattering between the selectively ionised U235 ions and the neutral U238 atoms which are present at 100 times the U235 abundance.

Other electro-magnetic process can be proposed (and the magnetic field associated with the steering and focussing of the electron beam heating is also present), but the preferred collection processes are closely guarded by the practitioners.

This description has concentrated on the physical principles of the process which may be new to this audience and the details of implementation have been glossed over most unfairly. Figure 9 shows a result obtained at Harwell some few years ago. The upper trace shows the alpha-particle spectrum from a thin layer of natural uranium. The alpha peaks from U235 are barely visible, but are entirely consistent with the 0.7% abundance of U235 in natural uranium and with the relative decay rates. The lower trace shows the alpha-particle spectrum of a similar layer of uranium, but this time after selective ionisation and collection of this sample. Comparison indicates an enrichment of the U235 component by a factor 9 - an early result, but showing that the potential exists to enrich natural uranium to the level required for civil power reactors. The continuing task is to determine whether at plant scale the process can be economically competitive with the gaseous diffusion and centrifuge processes.

6.   SUMMARY

This presentation has compared in broad terms the enrichment of isotopes of uranium on the large scale using Gaseous Diffusion (the oldest technology), the High Speed Centrifuge (the most recently developed process) and Atomic Route Laser Isotope Enrichment (a process receiving much attention to evaluate its commercial potential for the future). Attention has been drawn to some aspects of the processes which are similar in concept to those in Gas Separation with which the audience will be familiar.

Process Intensification: A Rotary Seawater Deaerator

V. Balasundaram and J. E. Porter

CHEMICAL ENGINEERING DEPARTMENT, NEWCASTLE UNIVERSITY,
NEWCASTLE UPON TYNE NE1 7RU, UK

C. Ramshaw

ICI C&P LTD, PO BOX 8, THE HEATH, RUNCORN, CHESHIRE WA7 4QD, UK

1 INTRODUCTION

Several years ago ICI developed a centrifugal distillation/absorption device (Ref 1) which gave a large hydraulic and mass transfer performance in a given rotor volume. This was part of a process intensification strategy for making major reductions in the size of process plant and hence in the cost, size and weight of processing systems. The key idea was the application of substantial accelerations to the countercurrent flow of gases and liquids in a fibrous torus having a high specific area and in which gas was the continuous phase. The intimate contact and large body forces generated at 10^3-10^4 m/s^2 gave equivalent plate heights for total reflux distillation in the region of 1.5 cms at liquid and gas mass fluxes of 25 kg/m^{2}s.

This technology has been licensed to Glitsch in the U.S.A. for further development and exploitation. It is clear that offshore oilfield operations represent an important niche market, because there, there is a particularly strong interest in reducing equipment size and weight. One of several interesting applications involves the deaeration of seawater for reinjection into declining oilfields in order to boost oil production. The oxygen concentration must be reduced to parts per billion levels in order to avoid biological contamination/blockage of the oil bearing rock and to avoid corrosion to pipework and this must be accomplished at water rates of 500 tonnes/hr or more.

In the original Higee concept with gas as the continuous phase, the radial liquid flow imposed two energy penalties. The first required the liquid to reach the peripheral rotor speed typically 60 m/s - and the second involved intense shearing of the thin liquid film as it drained outwards over the packing under the influence of the high body force. It can be shown that each of these energy components corresponds to $\frac{1}{2} \dot{m} Vt^2$ giving a total power demand of 500 KW for a water rate of 500 tonnes/hr. This was deemed to be unacceptable, even though the machine would have been much lighter and more compact than competing equipment at the time.

Further consideration of the contacting problem involving sparingly soluble gases in liquids leads to the observation that a relatively small volume of gas is needed to perform the stripping duty, compared with that required in a more soluble system. Thus for example the inert gas needed for oxygen stripping (at STP) is only 0.1 - 0.5 times the volume of water to be treated. If we follow a strategy of dispersing this gas in a continuous liquid phase it becomes feasible to return the liquid to an inner radial position as shown in Fig 1 where it can be discharged from the rotor with a greatly reduced velocity, typically 20 m/s. or less. The operation is then analogous to that of a centrifugal lute, with gas being injected on the periphery at a pressure somewhat greater than the local hydrostatic value P_T.

Assuming conservatively that liquid extends to the axis and that

$$\frac{\rho_\ell - \rho_g}{\rho_\ell} \approx 1 \text{ then } P_T = \frac{1}{2} \rho_\ell V_T^2$$

Hence for V_T= 60 m/s P_T= 18 bar.

For a water flow of 500 m^3/hr, and an air flow of 250 m^3/hr at STP, the three stage gas compression power is about 35 KW and the liquid kinetic energy at overflow is 28 KW. This gives a total power consumption of 63 KW compared with 500 KW required when gas is the continuous phase. This provides a considerable incentive for further consideration of this design strategy.

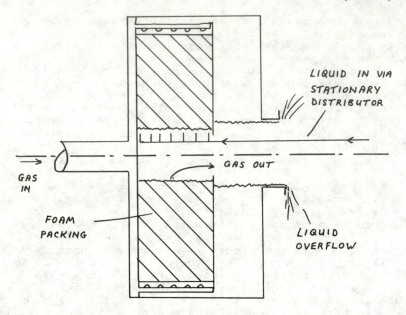

Figure 1 The proposed rotary deaerator with a continuous liquid phase

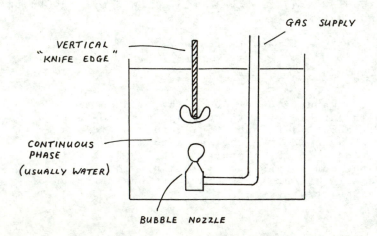

Figure 2 The impact cell

2   BUBBLE IMPACT STUDIES

In order that intimate and effective counter current gas liquid contact is achieved in the rotor, we need to generate a substantial volume fraction of fine bubbles. However the very high radial pressure gradients and the associated rapid bubble velocities tend to stimulate vigorous liquid phase mixing which is detrimental to efficient counter current operation. Therefore a rotor packing is needed which combines a very high surface area with a large voidage so that liquid circulation can be inhibited as much as possible while achieving an acceptable hold-up of fine bubbles.

An earlier study of liquid bubble behaviour in a packed liquid extraction column (Ref 2) is relevant to this problem. It was shown that when a bubble collided symmetrically with a vertical knife edge, its subsequent behaviour was dictated by the sum of the buoyancy and kinetic energy released on impact. If this was greater than the surface energy created by splitting then breakdown occurred, otherwise only deformation and bouncing could be expected. The simple energy equation predicted a critical bubble size above which breakdown occurred and this agreed remarkably well with that observed in impact cell experiments. A sketch of the impact cell and a typical sequence of photographs is shown in Figs 2, 3. More recently these simple experiments were repeated with air bubbles and once again a critical bubble size was observed. The size agreed well with the predictions of the energy equations, provided the bubble kinetic energy term included the virtual component associated with the liquid flow field around the bubble. Under terrestrial acceleration the critical size was about 3 mm.

It was further observed that horizontal wires or filaments were just as effective as vertical knife edges in causing bubble splitting (for super-critical bubbles), virtually irrespective of the fineness of the filaments. Multiple spaced layers of knitted mesh have been mounted in a bubble column to reduce the mean bubble size and eliminate the large bubbles responsible for vigorous local mixing (Fig 4). Hence it seemed reasonable to expect that a packing material consisting of a random bed of bonded fibres would act as an effective array of bubble-splitting elements, provided the bubbles could freely negotiate the packing voids. Since the specific area and voidage of the envisaged packings are about 3000 m^2/m^3, >90% respectively,

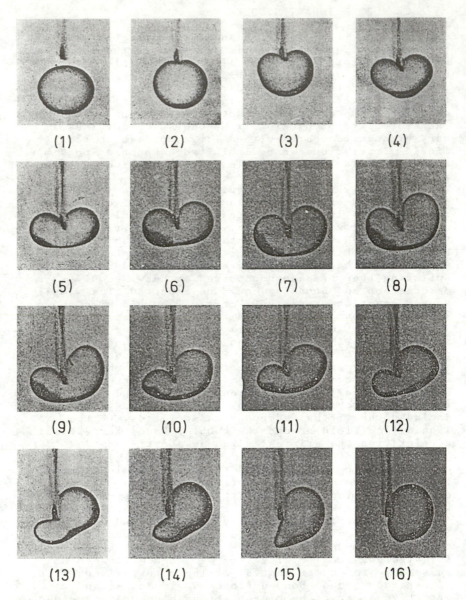

Figure 3a Impact sequence: subcritical bubbles

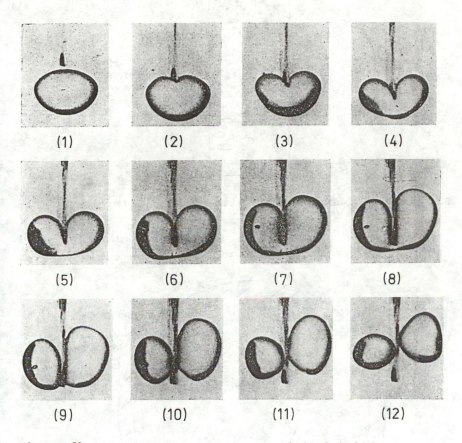

Figure 3b Impact sequence: supercritical bubbles

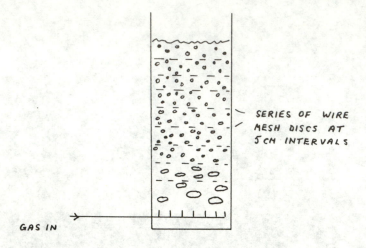

Figure 4   A bubble column with mesh splitters

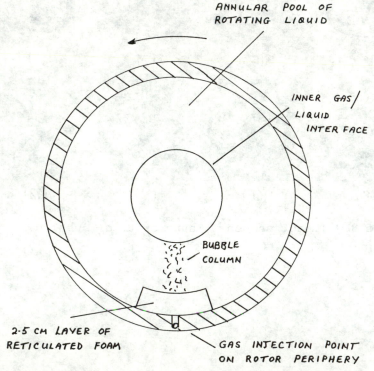

Figure 5   Bubble breakdown in a centrifugal field

bubbles much smaller than 3 mm are necessary to avoid
blinding the packing. The energy equation suggests that
at accelerations of 10^3 m/s^2 and over, 300μ bubbles may
be achievable. In order to explore this idea an
exploratory experiment was performed using the
apparatus shown in Fig 5.

Gas enters the shaft rotor through a union and is
conducted to the periphery through radial passages
terminating in a single orifice. The rotating annular
pool of water is about 1.2 cms thick and is bounded on
the front face by a Perspex sheet to permit easy
viewing of the column of gas which rises inwardly from
the orifice. Immediately above the orifice a block of
fibrous packing breaks up the gas into bubbles which
have been examined under various operating conditions
(Fig 6).

As anticipated, the packing is very effective in
breaking up the bubbles and becomes more so when the
rotational speed is raised. A further significant size
reduction is evident when tap water is replaced by
simulated seawater (tap water + 3.5% w/w NaCl). This is
due to the coalescence-suppressing action of the ionic
solute which should exert a beneficial influence on the
mass transfer intensity. However the lateral
dispersion of the bubble column in the packing is very
restricted, implying that quite a fine matrix of gas
orifices may be necessary to suppress radial
recirculation of the water and thereby achieve the true
countercurrent plug flow which is desired. The
incentive for doing this is outlined in Appendix 1
where it is shown that the predicted penalty caused by
backmixing/maldistribution is severe and becomes more
so with increased packing depth. It is therefore unwise
to assume that the number of transfer units is
proportional to packing depth until the liquid mixing
is largely suppressed.

Mathematical Model

As part of this study a computer model describing
the mass transfer performance of a packed rotor was
constructed. The model was based on design procedures
for conventional bubble columns with allowance made for
the different geometry. Mass transfer calculations
assumed uniform spherical bubbles having a local
diameter equal to the critical size corresponding to
the local acceleration. The calculated bubble terminal
velocity took into account the local bubble size and

Figure 6a   Bubbles in a centrifugal field (500 rpm)

Figure 6b Bubbles in a centrifugal field (1000 rpm)

Figure 6c Bubbles in a centrifugal field (1500 rpm)

acceleration, with mass transfer coefficients being based on this. Values for the local gas holdup and hence the gas-liquid interfacial area, were defined by the bubble rise velocity and the local gas and liquid flows.

TABLE 1   MODEL PREDICTIONS

Conditions		
	Liquid Flow	2.11 ℓ/s
	Gas Flow (STP)	1.5 ℓ/s
	Packing Depth	10 CM
	Axial Length	2.5 CM
	Outer Diameter	40 CM
	Rotor Speed	1000 RPM

A)   Unhindered Bubble Rise

Initial Bubble Size μ	NTU
225	2.75
180	4.00
153	4.9

B)   50% Hindered Rise Velocity

Initial Bubble Size μ	NTU
225	6.4
180	9.5
153	12.0

Model Predictions

Rotor performance estimates were calculated for the conditions outlined in Table 1 which represent a reasonably typical experimental arrangement. For the 10cm depth of packing about 2.75 transfer units may be expected. The bubble size ranged from 240μ at the periphery to 330μ at the inner liquid surface. The corresponding bubble terminal velocities were 1.02 m/s and 0.94 m/s This modest variation suggests that increased bubble size compensates for the reduced acceleration at small radii.

Table 1 summarises performance predictions with different assumed initial bubble sizes and degrees of hindered bubble rise. The latter exerts a very large influence via the increased holdup and interfacial area as the rise velocity is suppressed. Clearly a congested packing is preferred provided significant bubble coalescence is not thereby stimulated. As expected a smaller initial bubble size also produces better

performance. These predictions gave considerable
grounds for optimism provided the problems of machine
design and liquid mixing could be overcome.

Machine Design

In view of the encouraging results of the model
studies a machine was constructed to test the
performance predictions at a scale which was credible
in the light of the large water flows ultimately
required. A conventional packed column exploiting
enhanced accelerations transforms into a rotating
packed torus. The present machine Fig 7 incorporates a
torus with a radial depth of 10 cm and an axial
thickness of 2.5 cm rotating at upto 1500 RPM. This
could process up to 10 tonnes/hr water. The design
allowed for the axial thickness to be reduced to 2.5cm
to reduce the demand on liquid and gas. The stripping
gas used in the experiments was nitrogen which was fed
to the hollow rotor shaft via a rotary union.

The gas passed through two radial tubes into the
peripheral plenum chamber of the rotor. From this
chamber it was then fed into the outer periphery of the
annular liquid pool, using a membrane or some other
distribution arrangement. Gas bubbles generated and
discharged at the distributor under intense
acceleration travel radially inward under the influence
of very high buoyancy forces. They then
disengage/coalesce rapidly at the inner surface of the
liquid annulus.

Liquid is introduced to the rotor from a
stationary nozzle distributor close to the inner liquid
surface, and is discharged at the outer radius, thereby
providing a counter current flow regime. However,
having reached the packing outer radius, the liquid is
then transferred axially into another chamber where it
moves radially inwards to a discharge lip only
marginally "outboard" of the gas-liquid interface in
the packed chamber.

Oxygen Analysis

In the air water system, the transfer of oxygen is
controlled by the liquid film. Under the conditions of
the present experiments, a considerable excess of
nitrogen was used to strip the oxygen, thereby ensuring
that the interfacial oxygen concentration was

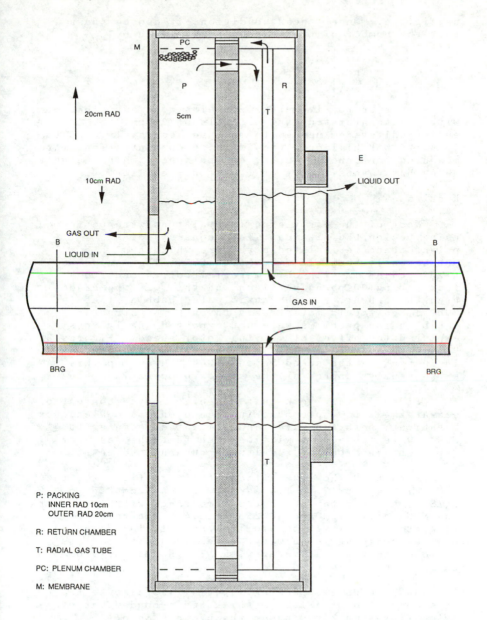

Figure 7   The machine design

negligible. Under these conditions it can be shown
that the number of transfer units is given by

$$\ln \left[\frac{C_1}{C_2} \right]$$

Since up to five transfer units were anticipated,
with inlet saturated oxygen levels of 9 ppm, the
analytical system needed to measure accurately down to
10 parts per billion. Liquid phase concentrations were
measured continuously using an Orbisphere membrane unit
which has a sensitivity of 1 part per billion.

Results

Most of the early experimental runs were performed
using Declon DEC 312 foam. This consists of a
polyurethane reticulated structure which has
subsequently been rigidised with PVC. The minimum cell
size obtainable with this material corresponds to a
foam having 20 pores per inch and this was the grade
used throughout. For the anticipated bubble sizes ($\approx$
200 μ), gas could move freely through the structure so
gas blinding problems were not expected. Early runs
explored the effect of varying rotor speed. Figures 8,
9 show that there is little improvement in performance
in increasing the speed beyond 1000 RPM and so this was
subsequently adopted as the standard operating value.

Since it was known that the electrolyte content of
seawater exerts a profound influence on its coalescence
behaviour, both tap water and simulated seawater (3.4%
NaCl) were tested. It will be seen that the seawater
gave very much better results throughout the tests.

Figs 8, 9 show the effect of liquid flowrate, the
NTU falling from 2 to 1.4 as flow rises from 0.6 to 2.4
kg/S with tap water. Over the same range with sea water
the NTU falls from around 2.8 to 2.4. The role of gas
flowrate (0.6 lits/S to 3.0 lits/sec) is indicated in
Fig 10. Once again seawater gives much better
performance than tap water with NTU increasing from 1.3
to 2.7 compared with 0.9 to 1.7.

An alternative packing material was also tested.
This consisted of a 1.2 cm thick resin-bonded mat of
nylon fibres marketed under the brand name of
Scotchbrite. The fibre diameter was approximately 50μ,
voidage 93% and specific surface area $\approx$ 3000 m^2/m^3.

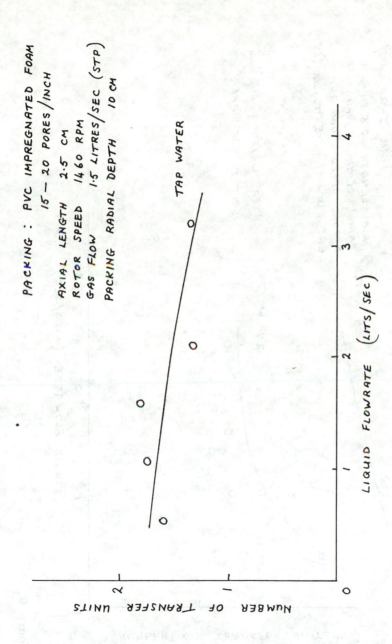

Figure 8

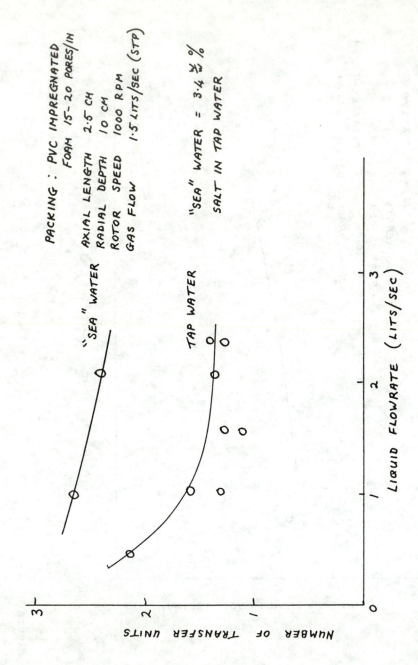

Figure 9

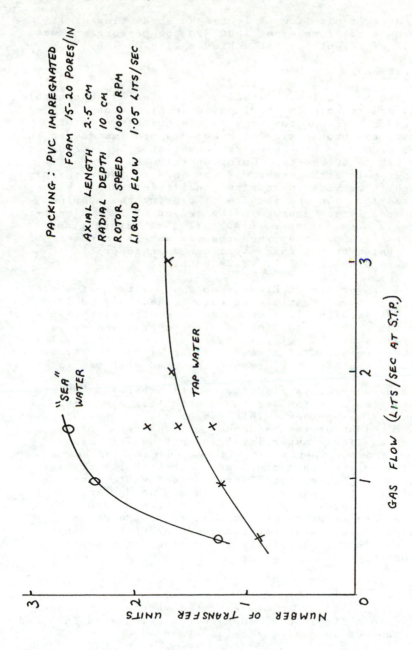

PACKING: PVC IMPREGNATED
FOAM 15-20 PORES/IN

AXIAL LENGTH 2.5 CM
RADIAL DEPTH 10 CM
ROTOR SPEED 1000 RPM
LIQUID FLOW 1.05 LITS/SEC

"SEA" WATER

TAP WATER

GAS FLOW (LITS/SEC AT S.T.P.)

NUMBER OF TRANSFER UNITS

Figure 10

At a liquid flow of 1.05 ℓ/s, a gas flow of 1.5 ℓ/s and a rotor speed of 1000 RPM the performance was little different to that achieved with Declon.

Discussion

The bubble impact studies show clearly that bubble sizes are reduced as the imposed acceleration increases. The photos suggest that within 5 cm of packing the bubble size is reduced approximately to the level predicted by the energy equation (150 microns at 1500 RPM, 15 cm radius). Unfortunately the simple photographic arrangement was not able to focus sharply enough to provide the desired quality of bubble size information. In the absence of packing the bubbles were much larger, being in the region of 1 mm diameter. There is some evidence to suggest that rather finer bubbles were produced when the gas was injected at sonic velocity but this may incur an unaceptable pressure drop in a full scale machine.

A comparison of Figs 8 and 9 suggests that there is little benefit in higher rotor speeds beyond 1000 RPM. It is also clear that the predicted performance with tap water is not forthcoming in practice - presumably because of its coalescence characteristics. On the other hand simulated seawater gives nearly twice the mass transfer performance observed with tap water.

The theoretical inverse relation between NTU and liquid flow is not observed, a much less sensitive behaviour being shown. This may be due to the increased bubble holdup or the reduced significance of back mixing as liquid superficial velocities are raised. Up to about 1.5 lits/sec, increased gas flow has a positive effect on performance, but with the present arrangement there seems little incentive to increase this rate. Perhaps the increased gas holdup and interfacial area is compensated by more vigorous back mixing. Experiments are in hand to test a denser packing in the outer rotor zone in order to inhibit recirulation and hopefully improve performance further.

Conclusions

1   Fine bubbles can be generated when gas is passed into a liquid pool containing reticulated foam and subjected to high accelerations.

2   A rotating torus of reticulated foam provides an intense environment for stripping/absorbing sparingly soluble gases from a liquid.

Acknowledgement

This work was supported by Glitsch Inc in a collaborative project with S.E.R.C.

Appendix 1.

The calculated effect of Backmixing on Mass Transfer Performance.

Consider a tower of unit cross section height h metres with a liquid flow L m/s having N transfer units. Inlet/outlet liquid oxygen concentration are C_1, C_2 respectively. We assume that the interfacial oxygen concentration is zero in view of the use of pure nitrogen for stripping and the poor oxygen solubility. Hence over an incremental height of tower dh we have (for plug flow of liquid)

$$K_L \ a \ dh \ c = -L \ dc$$

$$h = -\int_1^2 \frac{L \ dc}{K_L \ a \ c}$$

$$= \frac{L}{K_L a} \ln \left[\frac{C_1}{C_2} \right] \quad \text{Plug flow}$$

Hence, on the assumption of ideal plug flow of the liquid we evaluate the number of transfer units as

$$N = \ln \left[\frac{C_1}{C_2} \right] = \frac{h \ K_L a}{L} \qquad (1)$$

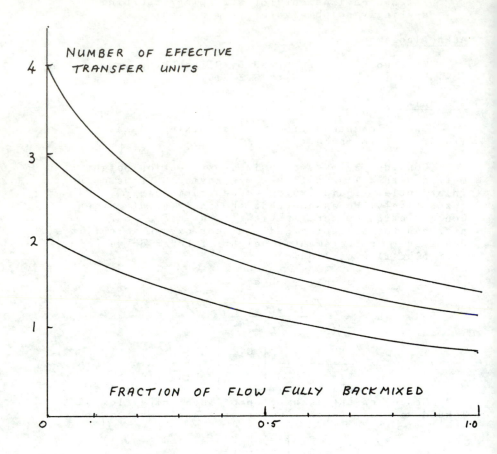

Figure 11   Predicted effect of backmixing on machine
performance

Where C_1, C_2 are the observed inlet and outlet concentrations respectively.

However the analysis is different if the liquid phase in the above column is well mixed. Now the volume is h m^3 so the oxygen mass balance gives

$$\frac{h K a C_2}{L} = L \quad (C_1 - C_2)$$

$$\frac{C_1}{C_2} = 1 + \frac{hK_L a}{L} \quad\quad (2)$$

If we now assume that the fully mixed and plug flow contactors are working in parallel with a fraction χ of the liquid flow passing through the fully mixed unit, then the exit average liquid composition is given by,

(C_2) Average $= \chi$ (C_2) fully mixed $+ (1-\chi)$ (C_2) plug flow

$$\frac{\chi C_1}{\left[1 + \frac{hK_L a}{L}\right]} + \frac{(1 - \chi) C_1}{\exp \frac{(hK_L a)}{L}}$$

When the experimental performance is assessed on the basis of ideal countercurrent flow then the effective number of transfer units is

$$\exp \left(\frac{C_1 \text{ average}}{C_2} \right) \quad\quad \text{Fig 11}$$

shows the relation between the effective number of transfer units and the fully mixed flow fraction.

REFERENCES

1. C Ramshaw, Chemical Engineer pp 3 Feb 1983

2. C Ramshaw, J D Thornton Instn Chem Engrs Symp Ser (Liq Extn) pp 73 April 1967

NOMENCLATURE

a	Specific surface area of packing	m^{-1}
c	Oxygen concentration in liquid	kg/m^3
h	Tower height	m
K_L	Mass transfer coefficient	m/s
L	liquid superficial velocity	m/s
$\dot{m}$	liquid mass flow	kg/s
N	Number of transfer units	-
P_T	Peripheral pressure	N/m^2
V_T	Peripheral velocity	m/s
ρ_ℓ	Liquid density	Kg/m^3
ρ_g	Gas density	Kg/m^3

Subscripts

1 at inlet

2 at outlet

The Role of Absorption in Gas Separation

Q. M. Siddique

BRITISH GAS PLC, RESEARCH AND TECHNOLOGY DIVISION, LONDON RESEARCH
STATION, MICHAEL ROAD, FULHAM, LONDON SW6 2AD, UK

Abstract — This paper gives a brief review of theoretical and
experimental aspects of absorption as applied to the separation of
acid gas impurities (CO_2 and H_2S) from fuel gases. The role of
absorbents (solvents), absorption equipment, relevant aspects of
process and plant design and data are discussed. Technological
problems, current practice and future trends are highlighted.

1   INTRODUCTION

Absorption is by far the most widely used technique for the
separation and purification of gases and is practised in chemical,
fertilisers, petrochemical, natural gas and substitute natural gas
industries. This paper deals with the separation (or removal) of
acid gases CO_2 and H_2S from fuel gas used in the gas industry.
Absorption involves the transfer of soluble components in a
gas—phase mixture into a liquid absorbent whose volatility is low
under process conditions. Desorption, or stripping, is the reverse
of the same operation in which the material transfers from liquid
to the gas phase. Methods used for gas separation using absorption
may be divided into three categories :

>   Physical solvents
>   Chemical solvents
>   Mixed physical plus chemical solvents.

There are references in literature to the earlier and
pioneering work from the gas industry by Silver and Hollings (1) on
design aspects of absorption. Cribb and Gibson (2) have studied
the absorption of CO_2 in ammoniacal solutions. In the general
literature there are very condensed surveys of the topic in, for
example, Chemical Engineers Handbook (3), in 'Absorption and
Extraction' (4) by Sherwood and Pigford, in Norman's 'Absorption
Distillation and Cooling Towers' (5), and elsewhere. Only recently

the knowledge on the subject has been put together by Kohl and
Reisenfeld (6), Astarita (7), Astarita Savage and Bisio (8).
Danckwerts (9), and Danckwerts and Sharma (10) have given a clear
exposition of the subject.

2 THEORETICAL BACKGROUND

The earliest and simplest theoretical attempt to describe the
mechanism of mass transfer in absorption was Whitman's two film
theory (11) which considered the resistance to mass transfer to be
in the region of gas/liquid interface. Later work showed that the
rates of mass transfer were not proportional to the diffusivity as
assumed in the 'film theory', but rather to the square root of the
diffusivity. Higbie (12) developed his penetration theory of
systematic surface renewal to account for this. Danckwerts'
'surface renewal theory' (13) is a modification of Higbie's. It
assumes no correlation between the time an element has been at the
interface and its likelihood of being transferred into the bulk.
For mass transfer involving chemical reaction in the liquid phase,
provided enough is known of the reaction kinetics, each theory can
be used to give an equation allowing for the effect of chemical
reaction. Danckwerts and Kennedy (14) showed that the rates of
mass transfer predicted by these theories for chemical absorption
systems were in close agreement. Danckwerts (9) has given a
comprehensive treatment of absorption followed by chemical
absorption and of different variations of the surface renewal
theory.

Figure 1 shows the place of CO_2 removal in the flowscheme for
substitute natural gas (SNG) production from naphtha or liquefied
petroleum gases (LPG) as practised in the gas industry in Britain
till the early 1970s. Purified naphtha after desulphurisation was
reacted with steam over a suitable CRG catalyst under optimum
condition of temperature and pressure. The gas-making process from
naphtha or LPG may be represented by a simplified equation as :

$$4(-CH_2) + 2H_2O \rightarrow 3CH_4 + CO_2 \qquad\qquad (1)$$

The resulting gas (mainly methane) contained about 21% CO_2,
which was then reduced to 2-4% (by absorption in activated hot
potassium carbonate solution) followed by drying, and enrichment by
propane or butane to correct for calorific value and Wobbe index
(defined as heating value of gas divided by square root of its
gravity relative to that of air) to achieve the required
specifications.

Another example (15) of gas separation by absorption refers to
the trial at the Westfield Development Centre of British Gas where
raw gas (obtained from the gasification of coal in British Gas Lurgi

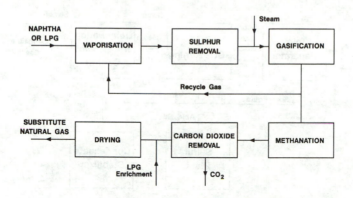

Figure 1 **CARBON DIOXIDE IN FLOWSCHEME FOR
SNG PRODUCTION FROM NAPHTHA OR LPG**

Gasifier) containing sulphur is purified to a sulphur content
(mainly H_2S) of less than one part per million by washing with
methanol. The purified gas is upgraded in the British Gas HICOM
process. In this process carbon monoxide and hydrogen gas are
converted to methane, the resulting gas containing about 58% CO_2.
To remove such a high level of CO_2, a versatile pilot plant has
been designed and built to investigate the efficiency of some
physical and chemical solvents for CO_2 and H_2S removal.

Figure 2 shows the place of acid gas separation in sour
natural gas plant.

3 SOLVENT (ABSORBENT) SELECTION

The solubility of gases in physical solvents is a function of
pressure and can be assumed to follow Henry's law. In such
solvents the gas to be separated (treated) is washed with a
suitable solvent under pressure, and the spent solution is then
regenerated by reduction in pressure with as little stripping as
possible. Use of physical solvents has many advantages,
particularly for the treatment of gases in which the partial
pressure of impurities is high and where the removal to low level
is not required. Examples of this type of solvent are propylene
carbonate (Fluor solvent process (16)), chilled methanol at about

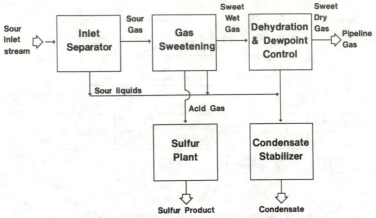

SOUR NATURAL GAS PROCESSING PLANT

−40°C (Rectisol process) (17) and dimethyl ether of polyethylene glycol (Selexol process) (18). Figure 3, as an example, shows acid gas separation flowsheet using physical absorption.

For bulk CO_2 and H_2S removal (separation) the older chemical solvents such as activated hot potassium carbonate and alkanolamines (ethanolamine, diethanolamine, di−isopropanolamine and diglycolamine) have been applied continuously although improvements include buffering, addition of reaction− rate enhancers, mixed amines, improved heat recovery and use of corrosion inhibitors.

Figure 4 shows a flow diagram of the activated hot potassium carbonate process for absorption of CO_2.

For selective H_2S removal, particularly at high pressure, physical solvents have undergone continuous development. Typical fuel gas applications with these features include natural gas processing, combined cycle (power generation from coal derived SNG) and SNG. One problem, however, to the application of solvents such as glycol ethers (Selexol, Sepasolv), methanol and N−methyl pyrrolidine (Purisol process) (19) is the high hydrocarbon co−absorption.

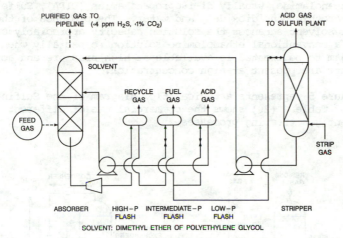

SOLVENT: DIMETHYL ETHER OF POLYETHYLENE GLYCOL

Figure 3 ACID GAS SEPARATION FLOWSHEET USING PHYSICAL ABSORPTION

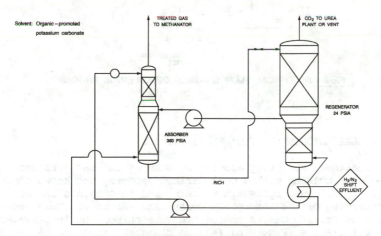

Figure 4 CARBON DIOXIDE REMOVAL SYSTEM USING CHEMICAL ABSORPTION

The Sulfinol process employs a mixture of chemical and physical solvents as the absorption medium. The solvent consists of an ethanolamine, usually di-isopropanolamine (DIPA), Sulfolane (tetrahydrothiophene dioxide), and water. The presence of the physical solvent enhances the solution capacity appreciably over that of a conventional ethanolamine solution, especially when the gas stream to be treated is available at high pressure and acidic components are present at high concentration.

Figure 5 represents a typical flow diagram of the Sulfinol process. Table 1 (20) shows the commercial use of different types of solvent gas treating processes.

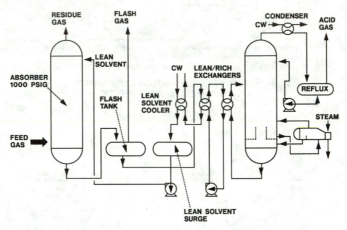

Figure 5   TYPICAL FLOW DIAGRAM OF SULFINOL PROCESS

4   LABORATORY AND PILOT PLANT STUDIES.

Idealised laboratory absorbers, such as laminar jets and wetted-wall columns with a well-defined geometry and known contact time for gas/liquid systems have been used to obtain a better understanding of the absorption mechanism. At London Research Station of British Gas, a medium pressure purpose built stainless steel pilot scale (Figure 6) (457 mm diameter) absorber has been used to study the absorption of CO_2 in diethanolamine-activated hot aqueous potassium carbonate solutions. The performance of the packed absorber over a wide range of operating conditions using various types of commercial packings was investigated. As

Table 1 (Ref 8) **COMMERCIAL USE OF DIFFERENT TYPES OF SOLVENT GAS TREATING PROCESSES**

Solvent	Number of installations
Acid gas removal	›1000
Aqueous alkanolamine	
MEA	
DEA	
DGA	
DIPA	
Promoted hot potassium carbonate	›740
Organic promoters	
Inorganic promoters	
Organic solvent – alkanolamine	›130
Sulfolane/DIPA	
MeOH/MEA/DEA	
Aqueous solution of potassium salt of amino acids	~100
Physical (organic) solvents	73
Propylene carbonate	
Polyethylene glycol dialkyl ether	
N – methylpyrrolidone	
Chilled methanol	

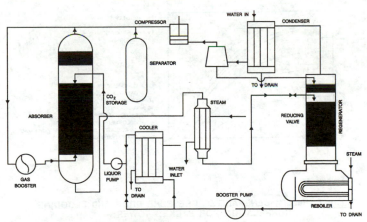

FIGURE 6 FLOW DIAGRAM OF PILOT PLANT FOR ACID GAS REMOVAL STUDIES

mentioned earlier, absorption of carbon dioxide by a solution of potassium carbonate is one of the standard cases illustrating chemical reaction in the liquid phase. In this type of absorption system the rate of mass transfer depends upon diffusion through the gas and liquid films and upon the rate of reaction in the liquid phase. Although diffusion and reaction in the liquid phase are believed controlling for the absorption of carbon dioxide in alkaline solution, the overall transfer coefficient $K_G a$ can be calculated from the following equation :

$$K_G a = \frac{N}{N\Delta_p} \qquad (2)$$

where $K_G a$ = overall mass transfer coefficient (kmol/m³/s/kpa)
       N = k mol per s (absorption of carbon dioxide in unit time)
       V = cubic metre (packed volume)
       Δ_p = bar (mean driving force)

For calculation, in—house equilibrium data were incorporated and for hydraulic checks the Generalised Flooding Correlation Figure 7 (21) was used.

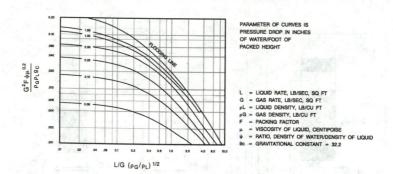

Figure 7 (Ref 21) GENERALIZED PRESSURE DROP AND FLOODING
CORRELATION FOR PACKED TOWERS

Figure (8) shows a plot of liquor flowrate vs K_Ga values. This curve tends to indicate that 2.5 cascade mini-ring packing gives higher values of K_Ga (mass transfer coefficient) with correspondingly lower pressure drop (Figure 9).

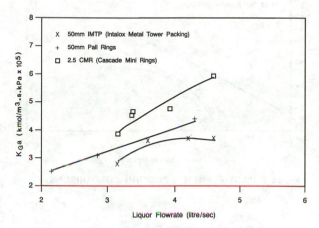

X 50mm IMTP (Intalox Metal Tower Packing)
+ 50mm Pall Rings
□ 2.5 CMR (Cascade Mini Rings)

K_Ga (kmol/m3.s.kPa x 105)

Liquor Flowrate (litre/sec)

Figure 8 PILOT PLANT 1 - PACKING COMPARISONS

Figure 10 shows the three packings used during this programme.

Laso and Bomio give a comprehensive review (22) on the design and modelling of structured packing (Figure 11) for selective absorption of H_2S from CO_2 containing gases. According to the authors the structured packing trade name "Mellapak" (see Figure 11), because of its high efficiency, high capacity, flexibility and low pressure drop has advantages leading to reduced absorber column dimensions and, at the high operating pressures often encountered, important savings in investment and weight. Weight is a critical factor for floating platforms and barges. Additionally, because of their higher capacity, structured packings are the packings of choice for debottlenecking existing plants. Figure 12 (22) shows a comparison of selectivity.

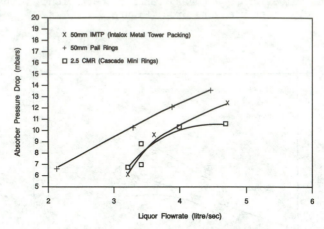

Figure 9 PILOT PLANT 1 – PACKING COMPARISONS

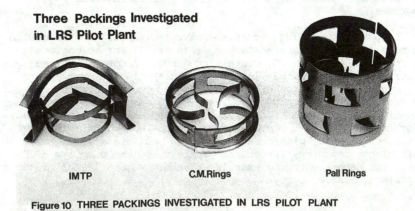

Figure 10 THREE PACKINGS INVESTIGATED IN LRS PILOT PLANT

Figure 11 STRUCTURED PACKING "MELLAPAK"

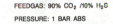

FEEDGAS: 90% CO_2 /10% H_2S
PRESSURE: 1 BAR ABS

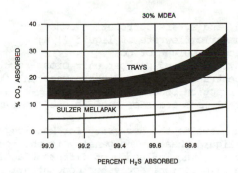

Figure 12 (Ref 22) **COMPARISON OF SELECTIVITY: MELLAPAK VS. TRAYS**

At present, structured packings of the type mentioned above are being evaluated for their performance in our pilot plant used for CO_2 absorption studies.

5 ROLE OF PILOT PLANT IN PROCESS AND PLANT DESIGN

It is important and desirable to have a pilot plant facility to obtain design and operating test data simulating as far as possible the full-scale conditions. Many of the major companies in the gas absorption field have appreciated the need for such a facility both on the technical grounds discussed above and from the standpoint of proving the process viability on smallscale. At Westfield Development Centre of British Gas a large scale (1350 Nm^3/h at pressure up to 70 bar) versatile pilot plant has been built and recently commissioned. This plant will provide design and operating data on the removal of CO_2 and H_2S from coal-based raw substitute natural gas (SNG).

6 MATHEMATICAL MODELLING

Siddique and Lam (23), after extensive laboratory scale and pilot plant work, put forward a design model for the absorption of CO_2 into activated hot aqueous potassium carbonate solutions. The heat, mass and rate equations have been solved incrementally through the packing.   The mass transfer coefficient has been regressed in the form of equation :

$$K_G a = \beta L^{a_1} M G^{a_2} \exp (a_3 C + a_4 yc) \qquad (3)$$

where

β, a_1, a_2, a_3 and a_4 are correlation constants
L   =   volumetric flowrate of liquid (l/s)
M   =   molecular weight of CO_2
G   =   mass flowrate of gas per unit cross-section (kg/m^2/h)
C   =   concentration of liquid phase reactant (kmol/l)
yc   =   mol fraction of carbon dioxide
$K_G a$ =   overall mass transfer coefficient ($kmol/m^3$/s/kPa)

This approach avoids the necessity for knowledge of reaction kinetics in the liquid phase. The penalty is the need for an extensive data-base. The influence of different activators is reflected in the calculation, activator concentration and fractional conversion of carbonate to bicarbonate.

This model has been confirmed with the plant tests and comparison of the calculated (predicted from the model) with the measured values of $K_G a$ (mass transfer coefficient)are given in Figure 13.

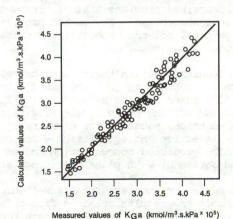

Figure 13 COMPARISON OF THE CALCULATED AND MEASURED K_Ga VALUES

Performance predictions produced with this model were compared with plant operating data. A good agreement (within 90% of the CO_2 removal) was obtained between the pilot plant data and model calculations.

O'Sullivan (24) describes a mathematical model for gas absorption by physical solvents (methanol in this case). The model is based on a set of approaches-to-equilibrium of the material and thermal performance of an absorber. A set of mass balance equations along with an overall energy equation are produced. The material balance equations are arranged as a tridiagonal matrix in the form of equation

$$\underline{A}\ \underline{X} = \underline{B} \tag{4}$$

Matrix $\underline{A}$ is of size $2n \times 2n$ containing coefficients relating the component molar flowrates contained in the vector $\underline{X}$ of size $2n$. Vector $\underline{B}$ contains the overall balance information.

The solution method adopted is that of Gaussian elimination (25).

The model has been used to simulate pilot scale measurements on the absorption of acid gases in methanol (24) in a ring—packed column. The model described provides a simple framework for the simulation of physical absorption. The method provides an adequate representation of physical solvent processes and thereby enables assessment of proposed flowsheets.

Hesselink and Spaninks (26) in their paper on selective removal of H_2S from CO_2 containing gas into mixed solvents have developed a model (27) based on mass balances for all absorbing species in both gas and solvent phase, point flux relationships for all species and an enthalphy equation for each tray. It is claimed that both kinetic — selective and equilibrium — selective operation can be simulated. The predicted concentration and temperature profile show a good agreement with plant operating data.

Mathematical models for countercurrent gas—liquid absorbers are important to enable the results obtained from pilot plants to be used for commercial design. A general approach has been to use stagewise models which are based on mass and enthalpy balances and incorporate general thermodynamic data. Examples of general purpose proprietary simulators (which include special data set for amines) are DESIGN II and PROCESS (from SIMSCI). AMSIM and TSWEET are examples of special purpose simulators for gas purification. They include simplified short—cut methods to estimate selectivity in tray columns.

The stagewise model (28,29), (30,31), however, is not the best one for the design of selective absorbers. Furthermore, if chemical absorption or kinetically—selective absorption are to be modelled, it is important to have a model that can deal with a chemical reaction in the liquid phase. Although there is a wealth of publications on absorption with chemical reaction in a liquid film, there are not many dealing with packed columns. The paper of De Leye and Froment (28) is one of the relatively few in which absorption followed by chemical reaction in packed columns is reported. Their model is a general one and relies on the splitting of the column in a relatively low number of sections which are handled as isothermal stages. Yu and Astarita (32,33) have presented a special purpose algorithm to simulate H_2S—CO_2 absorbers. Laso (34) has developed a special purpose column solution algorithm very similar to Yu and Astarita but including some numerical techniques which improve robustness and speed of convergence. A differential column model is being linked to the general purpose flowsheet simulator "PROCESS" so that the physical property databank and thermodynamic routines of the latter can be accessed. In this way, selective absorbers can be integrated in larger plants.

7 TECHNOLOGICAL PROBLEMS

Some of the major technological problems are :

Metallurgical,
Solvent degradation and
foaming of solvents

Astarita et al (8), and Kohl and Reisenfeld (35) provide a comprehensive review of the problem encountered commercially.

Metallurgical. The corrosion of carbon/mild steel by chemical gas treating solvents is an important issue for all types of chemical solvents. Corrosion affects the operating reliability and the expected life of the plant. In aqueous aminoalcohol solutions (used as absorbents), the allowable amine molarity depends strongly on the nature of amine. The reason is that it is the amine degradation products rather than the amine itself that promote severe corrosion, and amines differ as to their chemical stability and nature of their corrosion products. For example, MEA (Monoethanolamine) cannot be used in CO_2 absorption above 3M concentration without a corrosion inhibitor whereas DGA (Diglycolamine, which is also a primary amine) can be used up to 6M. Both aminoalcohols are degraded excessively by CO_2, yet DGA solutions are much less aggressive than MEA solutions to carbon steel. Equipment and procedures are available for measuring corrosion in amine plants. The use of both coupons and metal resistance corrosion probes is feasible. Corrosometer probes for plant use have been described. For CO_2 removal with aqueous MEA, one corrosion inhibitor consists of 4–6 parts by weight of antimony compounds and 6–4 parts by weight of vanadium compounds. Kohl and Reisenfeld (6) mention the use of 304 and 316 stainless steel for the hottest amine exchanger pass, the reboiler, the amine cooler, the amine reclaimer and certain sections of the piping.

Corrosion inhibitors are always needed when using potassium carbonate solutions for CO_2 removal. For simultaneous CO_2/H_2S absorption, the hydrosulphide ion acts to passivate the carbon steel, provided the H_2S concentration is not too low. Without inhibitors, hot potassium carbonate solution attacks mild steel severely. Probably the most effective corrosion inhibitor for use with potassium carbonate solutions is the arsenite ion. I hasten to add that, unfortunately, its use is severely restricted because of health, safety and environmental considerations. Vanadates are the most practical corrosion inhibitors for use with activated hot potassium carbonate solution. The active form for corrosion inhibition is potassium metavanadate (KVO_3) in which vanadium is in the pentavalent form. Vanadium is unusual in exhibiting five different valencies. Metavanadate may be reduced in solution to

the 4-valent form which is less effective, or ineffective, as a
corrosion inhibitor. In practice, the V^{5+}/V^{4+} ratio must be
maintained in a special range for good corrosion control.

Solvent Degradation.

Degradation is detrimental in that it leads to loss of active
aminoalcohol; it can also lead to operational problems such as
corrosion, foaminess, increase in solution viscosity and phase
separation. Means for removing degradation products are essential
for smooth operation of the plant. Proven techniques include
solution filtration, carbon treating, and thermal reclaiming. In
monoethanolamine degradation, the monoethanolamine carbonate (MEA
carbamate) is first converted to oxazolidone-2 which reacts with
one molecule of MEA yielding 1-(2-hydroxyethyl)-imidazolidone-2.
The substituted imizolidone hydrolyses to N-(2-hydroxyethyl)-
ethylene diamine and CO_2. It is possible to prevent the presence of
corrosion promoting diamine in the solution by removal of the
imidazolidone. The purification of aqueous monoethanolamine
solutions is effected by semicontinuous distillation.

Potassium carbonate is degraded by H_2S in the absence of CO_2,
the solution being converted to potassium bisulphide. Another
degradation path for K_2CO_3 solution occurs when treating gases that
contain carbon monoxide. Hydrogen and ammonia synthesis gases
contain low concentration of CO, whereas chemical synthesis gases
contain high CO concentrations, usually at high pressure. K_2CO_3
reacts with CO to produce formate ion and possibly oxalate ion :

$$CO + OH^- \rightarrow H-C{\overset{\displaystyle O}{\underset{\displaystyle O^-}{<}}}$$

Foaming.

Foaming may be caused by traces of dissolved impurities such
as degradation or corrosion products of the treating solution, fine
solids (e.g. corrosion products) or immiscible liquids (e.g.
condensed hydrocarbons). The theory of foam formation and
destabilisation has been discussed in literature (35). A foam may
be considered as a type of emulsion in which the minor phase is the
gas. It seems necessary to have some surface active component
present to stabilise the foam; pure liquids do not foam. All
liquid films tend to contract unless that tendency is counteracted
by the presence of a surface active component. The type of foam
encountered in chemical solvents contains mostly gas phase
surrounded by a thin film of solution. The mechanism of foam
collapse is film drainage by a process whose rate is governed by

what happens at the film boundaries. Foam lifetime depends on the
drainage rate. The properties of the liquid that increase foam
stability are high elasticity, high resilience and high surface
viscosity. The tendency of a liquid to foam, and foam stability
can be assessed in the laboratory by standard foam test apparatus.

The action of antifoam agents in preventing foaming can be
analysed in terms of the above parameters. It has been suggested
that effective antifoam agents must be liquids that are insoluble
in the foamy solution and have the ability to spread spontaneously
over the surface of the foamy liquid. This purely mechanical
action of spreading liquid ruptures the foam film. For inhibiting
foam in aqueous solution, the desired qualities in an antifoam
include insolubility in water, lower density than water, low
surface tension and low interfacial tension relative to an aqueous
solution.

Antifoams that have been found to be effective with aqueous
chemical solvents include hydrocarbon types such as polyethers and
silicone types such as polydimethylsiloxane containing free
particles of silica.

Foaming of amine solutions can be controlled by a carbon
filter which is used on a 10 per cent slipstream of the lean
solution at the pump discharge. UCON 5100 has been successfully
used as antifoam in activated hot potassium carbonate solutions for
CO_2 absorption.

8   CURRENT PRACTICE AND FUTURE TRENDS

Solvent screening programmes undertaken by various companies
interested in gas absorption as applied to gas separation have led
to the development of new solvent systems. For example, a new
hindered amine (36) for simultaneous removal of CO_2 and H_2S from
gases is claimed to give 50% increase in acid gas removal capacity
over promoted potassium carbonate treating. British Gas has
developed an amine promoted hot carbonate process LRS10 (37) for
CO_2 absorption which has been tested commercially with encouraging
and enhanced benefits. Porter, Sitthiosoth, and Jenkins (38)
describe the molecular design of a solvent for removing carbon
dioxide and hydrogen sulphide from gas mixtures. From basic
research on molecular interactions it now seems possible to
"tailor" and have so as to speak "Designer" solvent.

Structured packings as opposed to random or dumped packings
are finding frequent use in gas absorption for selective separation
of H_2S from CO_2/containing gases resulting in considerable savings
in absorber column size. Another particularly interesting
development aimed at reducing the absorber column size is the use

of increased gravitational fields to increase the liquid film mass transfer coefficient K_L. One example is ICI's Higee machine (39) in which the gas and liquid contactor is packing which spins at high speed. Such devices offer attractive advantages for offshore gas processing.

Database.

To define the driving forces to be used in calculations of mass transfer coefficients it is essential to have access to CO_2 and H_2S partial pressure data over a wide range of solvent composition. For physical solvents, typical data for binary systems are available in literature, but for multi-component gas mixtures, departure from ideality makes the predictive methods less accurate. Actual measurements are complicated and time consuming with the result that proprietary data are closely guarded. Figure 14 shows a block diagram as I see it for an acid gas removal scheme using the principle of absorption.

9   CONCLUDING REMARKS

Absorption, in spite of some impact and challenge from emerging technologies e.g. membranes and other adsorption processes, is still at the centre of the scene and plays a major role, and has a lion's share, in gas separation and will continue to do so for long time to come. In this paper I have touched upon various aspects of absorption for gas separation and it is fair to say that an intensive and extensive treatment of the subject falls outside the scope of such a meeting. There can be no doubt that the drive towards cost effective and environmentally acceptable solutions to gas separation problems will continue to receive the attention of researchers in this field. I emphasise with great concern that millions of tons of CO_2 released into the atmosphere worldwide contributes to the 'Greenhouse' effect. This tremendous quantity of CO_2 can be usefully consumed for enhanced oil recovery (EOR) or in urea manufacture for the fertiliser industry.

Note : Symbols and notations have their meanings given in the text.

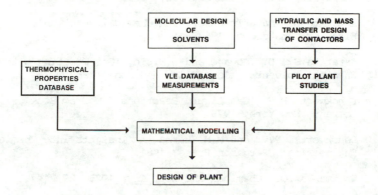

Figure 14 FUTURE RESEARCH IN GAS ABSORPTION

Acknowledgements : The author would like to thank Professor W.J. Thomas of University of Bath for this invitation to present this paper at the 5th BOC Priestley Conference, and British Gas plc for permission to publish this paper.

REFERENCES

1.   Silver L. and H.Hollings, <u>Trans. Inst. Chem. Engrs.</u>, London, 1934.

2.   Cribb G.S. and G.H.Gibson, <u>Trans. Inst. Chem. Engrs.</u>, 1964, <u>42</u>, T140

3.   Perry R.H. and C.H.Chilton, (ed.) 'Chemical Engineers Handbook' 5th Ed., McGraw Hill Book Company, 1973.

4.   Sherwood T.K. and R.L.Pigford, 'Absorption and Extraction', McGraw Hill Book Company, 1951.

5.   Norman W.S., 'Absorption, Distillation and Cooling Towers',
        Longmans, 1961.

6.   Kohl A.L. and F.C.Riesenfeld, 'Gas Purification', Gulf
        Publishing Company 4th Edition, 1985.

7.   Astarita G. 'Mass Transfer with Chemical Reaction',
        Elsevier, Amsterdam, 1967.

8.   Astarita G., D.W.Savage and A.Bisio, 'Gas Treating with
        Chemical Solvents', John Wiley and Sons, 1982.

9.   Danckwerts P.V. 'Gas Liquid Reactions', McGraw Book
        Company, New York, 1970.

10.   Danckwerts, P.V. and M.M.Sharma in 'Insights into Chemical
        Engineering', Pergamon Press, 1981.

11.   Whitman, W.G.   Chemical and Met. Eng; 1923, 29, 147.

12.   Higbie's Penetration Theory in P.V.Danckwerts' 'Gas-Liquid
        Reactions', McGraw Hill, New York, 1970.

13.   Danckwerts P.V., 'Gas-Liquid Reaction', McGraw Hill Book
        Company, New York, 1970.

14.   Danckwerts P.V. and A.M.Kennedy, 'Insights into Chemical
        Engineering', Pergamon Press, 1981.

15.   British Gas HQ News, 23rd August, 1989.

16.   Kohl,A.L. and P.A.Buckingham, Petr. Refiner 1960, 39,
        193-196.

17.   Herbert W., 1956 Erdol u. Kohle 9(2): 77-81.

18.   Sweny, J.W. "High CO_2-High H_2S Removal with Selexol
        Solvent" Annual GPA Conference, Houston, Texas, March,
        1980.

19.   Lurgi Gesellschaft fur Warme and Chemotechnik m.b.H Purisol
        for Gas Treating.

20.   Astarita G., D.W.Savage and A.Bisio, 'Gas Treating with
        Chemical Solvents',John Wiley and Sons, 1982 page 8.

21.   Eckert J.S., Chem. Eng. Progr, 1970, 66(3):39,

22.   Laso M. and P.Bomio, I.Chem.E. FSPG Symp. on "Removal of
        Sulphur Compounds from Gases", London October, 1988.

23. Siddique Q.M. and C.W.Lam, I.Chem.E, Jubilee Symp., London, 1982.

24. O'Sullivan P.A. Modelling of Gas Purification processes, paper presented at I.Chem.E. Annual Research Meeting London, 1988.

25. Davis M.E., 'Numerical Methods and Modelling for Chemical Engineers', John Wiley and Sons, 1984.

26. Hesselink W.H. and J.A.M.Spaninks, I.Chem.E. FSPG Symp., London, Oct.1988.

27. Spaninks, J.A.M., W.H.Hesselink and M.M.Svenson, AIChE Spring Meeting, New Orleans, March 6-9, 1988.

28. De Laye L. and G.E.Froment Computers and Chemical Engineering 1986 10(5).

29. Blauwhoff P.M.M., B.Kamphuis, Van Swaaij and K.R.Westerterp, Chemical Engineering Processes 1985, 19, 1-25.

30. Tomcej R.A.and F.D.Otto, Paper presented at the World Congress III of Chemical Engineering, Tokyo, 1986.

31. Ishii Y., Otto F.D., Can. J. Chem. Engrs. 1973, 51(5) 601.

32. Yu. W-C and G.Astarita, Chem. Engrg. Sci. 1987, 42(3), 425,

33. Yu. W-C and Astarita G., Chem. Engrg. Sci. 1987, 42(3), 419.

34. Laso M., Ph.D Thesis No. 8041, Swiss Federal Polytechnic Institute, Zurich, 1986.

35. Kohl A.L. and F.C. Riesenfeld, 'Gas Purification'. Gulf Publishing 4th Edition, 117-125.

36. Chem. Engrg. Prog. 1984, Vol 80, No.10.

37 Keene D.E., T.J.Ritter, P.A.O'Sullivan, GPA European/London Chapter 18th Nov., 1987.

38. Porter K.E., S.Sitthiosoth and J.D.Jenkins, I.Chem.E. FSGP Symp. London, Oct., 1988.

39. Starkey P.E. and B.Dobson, "Potential Offshore Applications for ICI Technology" I.Chem.E. Offshore Gas Technology Seminar, November 10, 1983.

Gas Separations through Reactions

M. M. Sharma and A. Mehra

DEPARTMENT OF CHEMICAL TECHNOLOGY, UNIVERSITY OF BOMBAY, MATUNGA,
BOMBAY 400 019, INDIA

1 INTRODUCTION

The recovery or removal of a variety of solutes, through
reactions, from diverse gaseous mixtures, continues to
occupy an important role in industry. Reactions are
sometimes deliberately imposed to realize higher rates of
absorption, provide much higher capacity per unit volume
of the absorbent and allow higher extent of separations
of the desired solute. In order that bulk separations are
economically attractive the reactions ought to be
reversible with temperature variations between absorption
and desorption often not greater than 100-120 °C. In many
cases the removal of a number of solutes, at low levels,
to vanishing levels, below even 1 ppm, is realized
through reactions only and in such cases the reactions
could well be irreversible. The manipulation of
chemistry, thermodynamics, kinetics and hydrodynamics
often permits selectivity with respect to the desired
solute, in the presence of other solutes. The range of
gas-liquid contactors is so broad that contact time can
be varied from a microsecond to a few seconds. Further in
continuous absorbers, with essentially plug flow, co-
current versus counter-current mode of contacting can
bring about considerable changes in selectivity when
simultaneous absorption of two or more solutes is
encountered.

The theory of gas absorption with reaction,
involving single or multiple solutes, one or more
dissolved reactants, irreversible or reversible
reactions, complex reactions, simultaneous absorption-
desorption, etc. has been developed sufficiently well. It
is often feasible to utilize data from carefully planned

laboratory experiments to approach the design of large plants, without recourse to pilot plant work; modern computational methods allow a more rigorous design procedure to be adopted.

In most cases in practice, alternative separation processes can be considered. However, absorption with reaction continues to have a niche in several areas.

2 FLUE GAS DESULFURIZATION (FGD)

FGD has acquired a lot of importance and a number of plants have been set up all over the world. The most common FGD process currently in operation is based on wet scrubbing with slurries of limestone or slaked lime as an absorbent and the product is calcium sulfate of high grade[1,2]. It is possible to produce calcium sulfite / bisulfite as a product if a ready market in the pulp industry exists. The FGD processes are highly capital intensive.

FGD plants are designed to handle dust laden gases and pressure drop is usually kept at a low level. A number of new designs,including the use of plastics as a material of construction, have been developed.

Role of Fine Particles

An analysis of different situations encountered in the wet scrubbing process, especially when the reactant particles are smaller than the diffusional scales of the diffusant (sulfur dioxide) in the liquid phase, brings out the role of fine 'sorbent' particles in a very effective manner.

Such analyses on the absorption of gases in a microslurry of sparingly soluble reactant were first reported by Ramachandran and Sharma[3]. A necessary condition for the intensification in the specific rate of absorption of SO_2 (species A) to occur is the depletion of the dissolved reactant (OH^- or CO_3^{2-}, species B) near the gas-liquid interface. This allows for a driving force for the release (dissolution) of B from the microparticles to be established. The situation for which this dissolution will prove to be of maximum benefit is the case of instantaneous reaction between A and B, since here depletion of B is maximum and very close to the interface the concentration of B is zero, thus leading to the highest possible driving force for dissolution.

The simple model proposed by Ramachandran and Sharma[3], under some reasonable assumptions, gives the following equation for the prediction of the specific rate of absorption:

$$R_A = (B_s/Z) \ [D_B k_{SL} a_p]^{1/2} \hspace{3cm} (1)$$

provided, $(D_B k_{SL} a_p)^{1/2}/k_L > 5$. Thus for small size particles, say, less than 5 μm, the specific rate becomes inversely proportional to the particle size. This finding is of great practical relevance and more so with slaked lime as on hydration of calcium oxide it is possible to manipulate conditions which do not require extra energy, to give very small particle sizes of calcium hydroxide approaching sub-micron dimensions.

Role of Additives

In some of the processes, such as the Saarberg-Holter process of Lodge-Cottrell, formic acid is added to the liquid phase, greatly increasing the solubility of the calcium by a factor as high as 800 and thereby enhancing the specific rate of absorption of SO_2. The formic acid maintains the pH of the solution between 4.2 and 5.2 causing the calcium formate to react with the dissolved SO_2 in the spray droplets to form calcium bisulfite rather than calcium sulfite. This bisulfite is then oxidized to calcium sulfate by sparged air (into the liquid sump) aided by a propeller mixer. The advantages of using formic acid thus include elimination of calcium sulfite deposits on the sides of the scrubber tower and because of the enhanced solubility of the calcium, the liquid to gas ratio may be reduced[2].

Role of Fine Bubbles

In some plants which are specifically designed to make sulfites for pulp plants, where three phase sparged reactors are used, we may encounter problems associated with fine bubbles whose characteristic dimension is less than the diffusion film thickness of gas-liquid and solid-liquid films. The existence of fine bubbles can make a marked difference to the specific rate of absorption of sulfur dioxide[4].

    3   SULFUR   RECOVERY FROM H_2S THROUGH LIQUID REDOX
         PROCESSES

The absorption of lean H_2S in the liquid redox sulfur recovery process is used to produce elemental sulfur.

These processes are used for a variety of purposes such as desulfurization of natural gas, refinery and chemical plant fuel gases, sour off-gas streams, vent streams from geothermal processes, coal gasification plants etc. It is possible to reduce H_2S to less than 10 ppm.

One set of processes is based on the classical Stretford process where vanadium is the primary catalyst interacting with the sulfur species. The Stretford process utilizes anthraquinone disulfonic acid (ADA) to catalyze oxygen transfer in the regeneration of the reduced vanadium. The kinetics of oxidation of a related hydroquinone is available in the literature[5]. Two subsequent modifications of this process, namely Sulfolin (Linde, A.G) - contains organic nitrogen compounds - and Unisulf (Unocal) -contains thiocyanate, carboxylic acid, aromatic complexing agent - have been reported[6].

In a second set of processes iron based chelates are used. For instance Lo-CAT (ARI) uses chelated iron compounds and so does the Sulferox process (Shell and Dow). The iron chelate based reaction results in the precipitation of sulfur and in a separate reaction ferrous is oxidized to ferric for recycle. The solution is regenerated by contact with air (pH of solution kept within 8.0-8.5 range). The advantages of the process are that it works at ambient conditions, has complete nontoxicity of solution and high catalytic activity, and above all it can be used at any pressure and without concern for gas composition because O_2, CO_2, NH_3, CO, mercaptans and most hydrocarbons have no ill effect on the solution[7].

In this context, an interesting system which shows 'autocatalytic' features has been reported by Mehra and Sharma[8] and is concerned with the absorption of hydrogen sulfide into aqueous solutions of iodine (solubilized by iodides). The variation of the specific rate of absorption with time in a batch (liquid) mode shows a maxima (contrary to what may be expected to be a conventional, decaying rate versus time trend). It was seen that the peak (maxima) value of the specific rate is about 8 times the initial value which approximately corresponds to the case of instantaneous reaction. A model which postulates instantaneous reaction between hydrogen sulfide and iodine in the liquid phase, precipitating product sulfur which acts as a microphase and transports the solute gas into the bulk liquid (see Section #5) and rapid second order agglomeration of the precipitating sulfur could explain the experimental results very well. Additives such as isopropanol, sodium sulfate were found to modify the agglomeration as well as the solubilizing characteristics of sulfur quite substantially. Thus the role of sulfur, when hydrogen sulfide is being absorbed and sulfur is a product, should be analyzed very carefully, given the considerable practical importance of redox based removal process where sulfur precipitates.

It is likely that in some situations in practice when ethane is treated for the removal of H_2S, with or without CO_2, supercritical conditions are encountered and such cases need to be analyzed carefully.

4 ABSORPTION OF NO_x

Selective Manufacture of Sodium Nitrite

The liquid phase reactions which result in the formation of nitrite and nitrate are as follows:

$$2\ NO_2\ +\ 2\ OH^-\ \longrightarrow\ NO_2^-\ +\ NO_3^-\ +\ H_2O \qquad (A)$$

$$N_2O_3\ +\ 2\ OH^-\ \longrightarrow\ 2\ NO_2^-\ +\ H_2O \qquad (B)$$

$$N_2O_4\ +\ 2\ OH^-\ \longrightarrow\ NO_2^-\ +\ NO_3^-\ +\ H_2O \qquad (C)$$

$$HNO_3\ +\ OH^-\ \longrightarrow\ NO_3^-\ +\ H_2O \qquad (D)$$

$$HNO_2\ +\ OH^-\ \longrightarrow\ NO_2^-\ +\ H_2O \qquad (E)$$

Nitrite is formed in reactions (B) and (E) and nitrate is formed in reaction (D) and both are simultaneously formed in reactions (A) and (C). For the selectivity towards nitrite, the formation of HNO_2 and N_2O_3 needs to be favoured whereas the formation of N_2O_4 and HNO_3 needs to be suppressed. The equilibria among the various species govern the individual concentrations. The concentration of N_2O_3 and HNO_2 increases with an increase in temperature and decrease in the partial pressure of NO_x. The concentration of HNO_2 increases with an increase in water partial pressure. Further, the concentration of N_2O_3 and HNO_2 increases with an increase in the ratio of divalent to tetravalent nitrogen oxides[9].

Gas phase mass transfer coefficient was found to play an important role in determining the selectivity. There was an optimum value at which the selectivity was maximum. The prowess of simulation, which involves the simultaneous solution of a strongly coupled set of differential species balances, has been elegantly brought out. In this case, the liability of dealing with lean NO_x, from nitric acid plants or from off-gases from nitric acid based oxidation of aromatic compounds plants and other sources, has been converted into a useful, reasonably priced, product as sodium nitrite[9].

Removal of Nitrogen Oxides from Flue Gases

A reversible absorption process for the removal of nitrogen oxides from the flue and off-gases, which could provide a viable alternative to the selective catalytic reduction process (of NO with NH_3) has been reported by Gestrich[10]. This process consists of absorbing NO in phosphoric acid esters and polyethylene glycol ethers which contain cupric halides.

Combined SO_2 and NO_x Removal

A new process uses an aqueous solution of hexamethylene tetraamine (HMTA) in combination with iron, zinc or aluminium. The basic idea is to enhance NO solubility in the SO_2 scrubbing solution by adding HMTA so that nitrosyl complex formation occurs[11].

5 USE OF MICROHETEROGENEOUS MEDIA IN ABSORPTION PROCESSES

Absorption of C_4 Olefins : Role of a Second Liquid Phase

The absorption of isobutylene in aqueous solutions of sulfuric acid is industrially practised to recover isobutylene from a variety of C_4 streams. Butene-1 and butene-2 (*cis* and *trans*) are also industrially absorbed in more concentrated sulfuric acid solutions for the manufacture of methyl ethyl ketone via *sec*-butanol. These absorption processes are capable of being influenced substantially in the presence of an emulsified, immiscible second liquid phase which shows a pronounced solubility for the olefinic solute gas compared to the original liquid (aqueous) phase. The emulsion normally has a dispersed constituent size much smaller than the diffusional length scales of the dissolved olefin in the liquid phase and hence an intensification in the specific absorption rate occurs due to the additional mode of transport of the solute gas via the emulsified phase (hereafter referred to as a microphase). The idea of using such a strategy was proposed by Sharma[12]; experimental data using chlorobenzene as a second liquid phase, along with an approximate theoretical analysis, were reported by Mehra and Sharma[13]. Very recently a comprehensive theory of such 'microphase catalysis' has been published[14,15]. The physical picture, in microphase catalysis, involves the uptake of the solute gas (A) by the microphase droplets near the gas-liquid interface. The microdroplets can rejuvenate themselves by any one or more of the following processes: subsequent circulation of these loaded microdroplets through the A-deficient (zero) bulk (where it sheds the solute) when the penetration element moves into the bulk liquid; removal of A through reaction in the droplet itself; and droplets escaping into the bulk liquid on account of their diffusive motions.

The simplified expressions for the specific rate of absorption, in presence of a non-reactive microphase, obtained are,

$$R_A = k_L A^*[(1-l_0) + l_0 m_A]^{1/2}, \qquad K_0 t_C/m_A \gg 1 \qquad (2)$$

$$R_A = A^*[D_A((1-l_0)k_1 + l_0 K_0)]^{1/2}, \quad K_0 t_C/m_A \ll 1 \qquad (3)$$

where, K_0 is the uptake coefficient which is equal to $12 D_A/d_p^2$ (assuming Sherwood No. = 2). Equation (2) describes a situation of local saturation of the microdroplets with respect to the solute A (small capacity and small drops) while eq.(3) assumes that the droplet acts as a sink (large capacity and large drops).

Extensive experimental data on gas-liquid (absorption of olefinic gases - isobutylene, butene-1, propylene - in emulsions and microemulsions of immiscible solvents such as chlorobenzene in aqueous solutions of sulfuric acid) systems have been reported by Mehra *et al.*[15]. The enhancements in the specific rate (defined as the ratio of rates in presence of the microphase to that in its absence) decrease as the value of the continuous phase rate constant is increased (by increasing sulfuric acid strength). The agreement between the predicted and the observed values was found to be excellent. The microphase hold up levels used in these studies ranged from 0.01 (v/v) to 0.20 (v/v) and enhancements as high as 25 (for butene-1) were obtained.

Separation of Gas Mixtures in presence of a Microphase

Consider absorption of species like styrene in off-gases at less than 1000 ppm, in aqueous solutions where reactions occur degrading styrene. The use of a second immiscible liquid phase where styrene is completely soluble can be very helpful. Similarly in the removal of malodorous gases typically present at less than 500 ppm, in aqueous hydrogen peroxide may benefit substantially from this strategy.

Sometimes even slower reacting gas removal can be made selective. The basic reason for realizing selectivity is associated with high distribution coefficient of the desired solute in the microphase. Tinge *et al.*[16] have reported the absorption selectivity between propane and ethylene in aqueous slurries of fine activated carbon particles. Mehra and Sharma[17] have considered this problem theoretically. The following equations give the selectivity factor in an oil-in-water and a water-in-oil emulsion, respectively,

$$E_s = [(m_A/m_B)/(k_{1A}/k_{1B})]^{1/2} \tag{4}$$

$$E_s = (m_A/m_B)/(k_{1A}/k_{1B})^{1/2} \tag{5}$$

the selectivity factor being the ratio of the enhancements in the specific rates of absorption of A and B. It may be seen that water-in-oil media may be relatively superior for enhancing the selective absorption of a desired component, since the distribution coefficient ratio now appears without the square root sign[18].

Some values of the selectivity factor that may be obtained from parameters for some typical systems using a non-reactive microphase such as an emulsified oil in a reactive aqueous phase are 1.8, 28.7 and 31.6 for butene-1 / propylene, carbon disulfide / carbonyl sulfide and carbonyl sulfide/carbon dioxide, respectively[17].

Absorption of Hydrogen in Slurries of Fine Particles of Metal Alloy Hydrides

The affinity of $LaNi_5$ alloy hydrides for hydrogen gas may be used for selectively removing hydrogen from its mixtures with other gases such as methane. Holstvoogd *et al.*[19] and Tung *et al.*[20] have reported data on the absorption of hydrogen gas into microslurries of these hydride particles in solvents such as silicone oil. Methane, on the other hand, shows no affinity for these alloy particles. These systems have relevance as hydrogen storage devices and may find use in enhancing rates of hydrogen transport limited hydrogenations as well as in achieving very efficient removal and recovery of hydrogen from a variety of streams. The extent to which hydrogen can be loaded is quite high and could approach 10 % by weight of the alloy particles. Further, the existence of particles having a size smaller than the diffusion film thickness can result in an enhancement in the specific rate of absorption in a manner analogous to that for an emulsified second liquid phase. The mathematical analysis of such situations is however more difficult since the distribution equilibria for hydrogen (alloy/oil) is highly non-linear in nature.

Cyclodextrins (CD): a Uniform Microheterogeneous Media

Cyclodextrins dissolved in water provide a novel media for conducting separations since they comprise of cage-like macromolecules dispersed in a continuous phase and these can form complexes with solutes. The removal of vapours of waste, volatile solvents generated in a variety of processes, may be processed for the recovery of the organic substance via absorption in a CD solution at a temperature at which the complex crystallizes which can then be removed by centrifugation and the CD solution recycled after thermal regeneration[21]. A rigorous analysis of such processes does not seem to have been carried out.

6 ABSORPTION WITH REVERSIBLE COMPLEXATION/REACTION

Absorption of Carbon Monoxide

The selective absorption of carbon monoxide may be accomplished by the use of metal complexes. Sato *et al.*[22] have reported the equilibrium and kinetic features of the selective and reversible complexation of carbon monoxide with copper (I) tetrachloroaluminate (III) in an aromatic hydrocarbon (COSORB process) such as benzene or toluene or xylene. The uptake capability of the copper complex is approximately proportional to its concentration whereas the solubility of carbon monoxide decreases with temperature and with total concentration of the aluminate.

The kinetics of absorption of CO, in the absence of acid gases like CO_2, H_2S etc., in aqueous solutions of sodium hydroxide and aqueous calcium hydroxide slurries has been recently reported by Patwardhan and Sharma[23]. A stirred autoclave with a flat gas-liquid interface was used to study the absorption process in a temperature range of 373-433 K and a pressure domain from 1 to 10 MPa.   Whereas with sodium hydroxide the CO reaction is substantially fast enough to be diffusion controlled, for the calcium hydroxide slurry it occurs partly in the bulk liquid phase. This process is useful for making sodium / calcium formate.

The selective absorption of CO, in the absence of any CO_2, from a variety of gases containing H_2, N_2 etc. in methanol containing potassium methoxide to give methyl formate (MF) is by now well known. This is not only a method of manufacturing MF but also of pure CO as MF can be conveniently decomposed to methanol and CO where the former can be recycled. The kinetics of absorption of CO in methanol containing potassium methoxide has been studied[24].

Olefin Separation

It is well known that complexing solutions such as cuprous tetrachloroaluminate in aromatic solvents can affect the recovery of ethylene from gaseous streams, e.g. gas streams from fluid catalytic cracking units and purge gases from ethylene oxide, ethylene glycol etc. This process continues to attract attention and cuprous aluminium chloride dissolved in toluene as an absorbent has been modified with aniline / toluidine[25].

The separation of C_2-C_5 olefins from paraffins and linear α-olefins from internal and branched olefins of the same carbon number may also be affected by a solution of cuprous hexafluoroacetylacetonate (diketonate) in a weakly complexing solvent for instance. The solvent, which is usually an olefinic hydrocarbon, stabilizes the diketonate against disproportionation of Cu(I) to Cu(II) and metallic copper. The separation of ethylene / ethane, propylene / propane, butene-1 and other butenes as well as 1-pentene and other pentenes using the diketonate has been discussed by Ho et al.[26]. An important observation is that the olefin uptake by complexation increases with increasing steric hindrance at the double bond of the solvent. Thus the highest ethylene uptake was found to occur when α - methyl styrene was used (highest degree of steric hindrance). The basic fact responsible for separation is the higher stability of linear α-olefin complexes with the diketonate as compared to the internal and branched olefins.

Certain complexes may prove to be versatile for more intricate separations such as the recovery of ethylene in presence of carbon monoxide. This type of selective absorption is shown by a toluene solution of aluminium

silver chloride[27] at 293 K and atmospheric pressure. No
CO is absorbed at these conditions. This is because it
cannot effectively repel the toluene from the Ag(I) ion
due to weaker coordinating ability than ethylene.

The selective absorption of isobutylene, in the
presence of butenes (butene-1 and *cis* and *trans* butene-2),
has been a classical example for over forty years,
for obtaining even polymer grade material. The new
strategy in this area has been the selective reaction of
isobutylene with methanol, in the presence of strongly
acidic ion exchange resin, to give methyl *tert*-butyl
ether (MTBE) which on cracking gives pure isobutylene of
polymer grade. Thus this not only becomes a method of
manufacturing MTBE but also a process to make pure
isobutylene. This strategy also works for the selective
removal of isoamylene from the C_5 olefin mixtures
containing pentenes, isoamylenes etc. In place of
methanol, ethanol can also be used and even ethylene
glycol has been claimed to be an effective reactant[28].
Recent work of Patwardhan and Sharma[29] shows that the
mixture of butenes and isobutylene can be reacted with
acetic acid and at higher temperatures, around 373 to 393
K, with cationic ion exchange resin like Amberlyst 15, to
allow selective reaction of butene-1 and -2 and
isobutylene does not react. By contrast, at lower
temperatures, around 278 to 283 K isobutylene reacts with
high selectivity and butenes hardly react.

7   RECOVERY FROM LEAN STREAMS

Recovery of Dry HCl from Lean Mixtures and Separation of HCl / HBr

HCl reacts with weak aromatic amines like aniline /
toluidines / cumidine etc. in non-polar solvents like
toluene, ethyl benzene, cumene etc. Here the amine-
hydrochloride comes out as a separate phase which can be
removed and taken separately in the above solvent and
boiled when HCl comes out. The reaction of HCl with the
amine is instantaneous.

The difference in the acidity of of HCl and HBr can
be exploited for separation and recovery by absorption in
an amine solution[30].

Absorption of Lean Chlorine

Lean mixtures of chlorine are encountered in several
industries including the 'sniff gas' in caustic-chlorine

plants. The absorption of Cl_2 in aqueous alkaline solutions is a convenient method for removing Cl_2. A new approach to this problem has been provided by Lahiri et al.[31] where absorption is carried out in hot aqueous NaOH and gas side and liquid side mass transfer coefficients are suitably manipulated to allow selective desorption of pure HClO which in turn allows making pure NaClO by separate absorption in NaOH. The theoretical aspects of this problem have also been covered.

Bromine from Lean Bromide Solutions

Lean mixtures of air and bromine, resulting from air stripping of bromine from chlorinated bromide liquors, require bromine to be recovered. Apart from the standard method, where bromine is absorbed into aqueous alkali or carbonate solutions, some novel approaches have been reported. One of them consists of absorption of bromine in an aqueous solution of some specific quaternary ammonium bromide where a water immiscible compound is formed and can be easily dissociated to liberate bromine. Compounds such as N-ethyl pyrrolidinium bromide and N-ethyl morpholinium bromide have been patented[32].

8 ABSORPTION IN MELTS

Dunbobbin and Brown[33] have reported a promising method for producing 99.8 % pure oxygen from air. A molten mixture consisting mainly of alkali nitrates and nitrites at 898 K reacts chemically with oxygen in compressed air. When the molten, oxidized mixture is depressurized and/or heated, it gives off oxygen as a pure gas via a reversible mechanism. The kinetics of this reaction has not been reported in the open literature. This process appears to be attractive and if successful will show the prowess of absorption processes in an impressive way.

9 IMPURITY REMOVAL

Organic Impurities Removal

With the development of extremely stringent standards for environmental control, not only is the need for controlling emission levels of undesirable compounds more urgent, economic pressures suggest a shift from completely destructive techniques such as incineration to recovery based methods which may enable some offsetting of the extra pollution control costs.

A useful example is the removal of maleic anhydride, along with some phthalic anhydride, from the off-gases of phthalic anhydride (based on o-xylene) plants. Here absorption in water converts maleic anhydride into water soluble maleic acid and the absorbent can be in a recycle system with a bleed which can then be processed to be converted by a simple homogeneously catalyzed isomerization to sparingly soluble fumaric acid which can be separated and the solution recycled.

Removal of Carbonyl Sulfide (COS)

Some amount of COS is invariably present in natural gas streams and C_3 streams originating from petroleum processing units whereas sulfur is present in the raw feedstock. The kinetics of COS absorption in aqueous caustic alkaline solutions as well as aqueous solutions of alkanolamines were reported by Sharma[34]. The intrinsic rate of reaction of COS with any amine is roughly 100 times lower compared with CO_2 whereas for OH^- this factor is about 10^3. Higher selectivities for COS may be obtained when present along with CO_2 by using ethanol in the aqueous phase which increases the solubility of COS by a much larger factor than for CO_2.

Chaudhuri and Sharma[35] have recently reported studies on the absorption of COS in emulsions of toluene in aqueous alkali solutions (microheterogeneous media). Enhancement factors in the specific rate of absorption, as high as 2.5 have been observed.

The removal of COS from comparable boiling point hydrocarbons propane / propylene may be advantageously converted from the gas-liquid to the liquid-liquid mode as pressures are around 2 MPa as in the latter case interfacial areas can be relatively very large and even phase transfer catalyst can be employed.

Phosphine Removal

Acetylene derived from calcium carbide invariably contains phosphine as an impurity which may range from a few ppm to 2000 ppm. Phosphine is also generated in the production of phosphorus by the electrothermal route as well as in the manufacture of calcium hypophosphite.

Chandrasekaran and Sharma[36] have reported the kinetics of phosphine absorption in aqueous sodium hypochlorite at different pH values. The desirability of conducting reaction at lower pH (9.5-9.8) is clearly demonstrated, compared to a higher pH of 13, since the rate constant value in the lower pH range is much higher. The data of these authors covers the range of industrial importance. In all cases relating to industrial contactors the system conforms to the case of absorption accompanied by fast reaction, for a pH less than 12.5.

Sulfuric acid of 80-92 % strength may also be used to absorb phosphine, wherein the absorption is accompanied by a pseudo first order reaction.

Removal of Silanes

Silanes exhausted from production processes for crystal growth of silicon in the electronic industry need to be completely removed since they pose tremendous health hazards. Sada *et al.*[37] have reported the kinetics of absorption of silane accompanied by its alkaline hydrolysis and they found the hydrolysis reaction to the first order with respect to both the dissolved silane as well as the sodium hydroxide.

Mercury Removal

This is an unusual example and is encountered in a variety of industries such as the electrolytic caustic soda / chlorine / mercury cells, off-gases from smelters etc. The acute toxic nature of mercury poses serious problems and hence it must be removed from the relevant effluent streams. Absorption of mercury into aqueous hypochlorite solutions is a promising method. The hypochlorous acid reacts with mercury to form mercuric oxide (and chloride ion) which may further undergo reaction with the excess chloride ions and water to give a tetrachloro complex. Nene[38] has shown that the absorption of mercury from air in aqueous hypochlorite solutions (containing different concentrations of electrolytes) is accompanied by fast pseudo first order reaction in the liquid phase; reaction rate constants are around 1×10^5 $m^3/kmol.s$.

10 REACTIVE MEMBRANES

The use of membranes for affecting separations is an area of great interest in view of some inherent advantages offered by this system.

Since diffusion in liquids is much faster than say, in polymer films, fast separations with high selectivity may be achieved using liquid membranes. The selectivity in membrane systems is due to the use of a mobile carrier which facilitates, by reversible complexation, the transport of a desired species exclusively. The use of liquid membranes industrially, however, poses problems, primarily those of instability of the membranes due to dissolution in surrounding solutions, osmotic pressure, leakage of inner materials, etc.

A class of membranes which utilize the full potential of liquid membranes are known as immobilized liquid membranes (ILMs). It is possible to develop ultrathin ILMs using a variety of techniques. This is necessary because permeation rates through conventional ILMs are rather low. For instance a composite hydrophilic-hydrophobic membrane may be formed with a thin hydrophilic skin wherein aqueous solutions can be immobilized. Sirkar and Bhave[39] have reported a novel alternative strategy to prepare ultrathin ILMs in a single ply membrane structure. Thus membranes are obtained with liquid immobilized in the microporous hydrophobic polymer support. The use of ion exchange membranes as supports for immobilizing liquids has the potential of eliminating the disadvantages of using a microporous support[40]. An immobilized, microheterogeneous phase, such as an emulsified second liquid phase,may considerably increase the flux of the solute through an ILM in a manner analogous to that described in Section #5. Teramoto et al.[41] have reported the separation of ethylene from ethane by supported liquid membranes containing silver nitrate as a carrier. Very recently these workers have used a flowing liquid membrane solution, for the same system, where the arrangement consists of a liquid carrier solution flowing in a thin channel between two microporous membranes[42].

The transport of oxygen through thin layers of haemoglobin solutions supported by cellulose acetate fibres is another typical case and measurements of the steady state flux across these membranes were reported by Scholander[43].

Oxygen enrichment may be affected by a cobalt complex incorporated in a membrane. Cobalt (II) complex with poly (ethylene imine) forms a reversible, stable - peroxo adduct in aqueous solutions at room temperature. The value of the oxygen release rate constant is fairly high and therefore allows for convenient discharge of oxygen. An increase in the degree of crosslinking of the membrane as well as nitrogen content decrease the O_2 binding ability of the membrane[44].

A temperature swing process, which utilizes a hollow fibre reactor using a suspension of an encapsulated manganese complex in water, has been reported and is capable of generating pure oxygen at near ambient conditions[45].

Air products (USA) have claimed that CO and N_2 can be separated with a selectivity of 30 to 40 by contacting the mixture with a Celgard 3501 film loaded with liquid ET_3NH^+ [Cu_2Cl_2] salt. In a similar manner CO_2 and CH_4 can be separated by a film of tetrahexyl ammonium benzoate melt encased between 2 sheets of gas permeable poly (n-methylsilylpropyne) with an achieved selectivity of 45[46].

The technique of using melts is relatively novel with operating temperatures of about 573 K. The molten salts can reversibly complex with specific gases such as CO_2, NH_3 and act as a barrier to other gases like H_2, He[40]. The potential of adopting this methodology in ammonia plants is high and can bring out important changes in the flowsheet.

A promising way of improving membrane stability is to bind a mobile carrier chemically within the membrane. Sharma[12] has referred to the possibility of using membranes containing reactive groups as an integral part of the polymeric membrane and cited, for instance, the example of polyethylene imine membrane for removal of acid gases. Very recently Cussler et al.[47] have reported a theory which describes transport across a membrane bearing such chained carriers. A chained carrier membrane may be looked at as a lamellar structure where the carrier is located in layers each of thickness p. Each carrier can move over a distance p_o around its equilibrium position but cannot move permanently into another layer. This restricted motion of a carrier thus allows a complexed solute to be transferred to an adjoining carrier, provided that $p < p_o$ i.e., there is some overlap between adjacent layers. This condition can be achieved only at some minimum carrier loading, below which no solute can be transported across the membrane. Cussler et al.[47] have discussed the theoretical aspects of this problem. The apparent diffusion coefficient for transport across the membrane, under certain simplifying conditions is given by, $D_{app} = \{k\underline{c}p^3(p_o-p)/p_o^2\}$. For $p > p_o$ this coefficient is zero and this characteristic enables us to distinguish between facilitated transport through a mobile carrier and a chained one.

An interesting example of facilitated transport through membranes with fixed carriers is provided by the permeation of O_2 through wine coloured membranes containing a series of cobalt and iron tetrakis (alkyl-amidophenyl) porphyrin derivatives as reported by Nishide and Tsuchida[48]. These membranes are prepared by covalently bonding or dispersing the metallo porphyrin with / into poly (alkyl methacrylate), poly (dimethyl siloxane) or poly (tri methyl silyl propyne). The permeability ratio for air components is 10 (pO_2/pN_2) and the facilitated transport occurs because these polymers bind molecular oxygen specifically, rapidly and reversibly even after fixation in solid state. The various kinetic constants were determined spectroscopically and it has been observed experimentally that the facilitated transport is more enhanced at lower oxygen solubility; also the diffusion coefficient via the fixed carriers increases with the oxygen dissociation kinetic constant from the fixed carrier.

11 GAS-LIQUID CONTACTORS

Doraiswamy and Sharma[49] have discussed many relevant aspects of gas-liquid contactors for reactions. The advantages of co-current versus counter-current mode of contacting in packed towers, low versus high mass transfer coefficient contactors, etc. have been discussed at length. For irreversible reactions there are clear merits in using packed columns with the co-current mode of contacting. The usefulness of packed towers has improved with better design of liquid distributors, support plates and packings made out of high performance plastics like PVDF, PTFE, etc.

The absorption of SO_2 and NH_3 from air and air/CO_2 mixtures has been studied by Cooney and Jackson[50] in a novel hollow fibre device. The gas and the liquid were made to flow countercurrently and it was found that the membrane provides a relatively small resistance to the interphase transport (especially at high liquid rates). The major advantages lie in avoiding reduction in interfacial area at low liquid rates as well as flooding at high liquid rates which are common in conventional towers. Alkaline liquids provide for very slow buildup of the dissolved SO_2 such that a large driving force can be maintained for the mass transfer. According to Qi and Cussler[51] the membrane resistance may become significant if the liquid wets the membrane. They have shown, however, that the advantage in terms of increased area for transport supersedes the disadvantage of higher overall resistance to mass transfer because of the membrane.

12 CONCLUSIONS

Gas separation through reactions continues to be an important and challenging area. Further advances in flue gas desulfurization, particularly combining SO_2/NO_x removal, can be expected. The use of novel complexing agents which react with solute gases holds a promising future. The manipulation of thermodynamics, kinetics and hydrodynamics may promote selectivity for the desired solute. The use of microheterogeneous media can bestow several advantages in not only enhancing the rates of absorption but also realizing selectivity. Hollow fibres for membrane based separations as well as contactors for gas-liquid reactions are expected to become increasingly important.

NOTATION

A^*	solubility of solute gas in liquid phase, $kmol/m^3$
a_p	specific interfacial area of fine particles, m^2/m^3 (slurry volume)
B_s	solubility of solid reactant in liquid phase, $kmol/m^3$
$\underline{c}$	average carrier concentration, $kmol/m^3$
D	diffusivity of solute, m^2/s
d_p	diameter of microparticles, m
E_s	selectivity factor
K_o	microphase uptake coefficient for solute, s^{-1}
k	exchange reaction constant for solute between adjacent sites, $m^3/kmol.s$
k_1	pseudo first order rate constant, s^{-1}
k_L	liquid side mass transfer coefficient, m/s
k_{SL}	mass transfer coefficient for fine particles, m/s
l_o	fractional volumetric hold up of microphase
m	distribution coefficient of a solute (microphase / continuous phase
p	layer thickness, m
p_o	distance of chained carrier mobility, m
R_A	specific rate of absorption, $kmol/m^2.s$
t_c	surface element contact time, s (Higbie's mass transfer model: $= 4D_A/k_L^2$)
z	stoichiometric factor for reaction between A and B : A + ZB -> products
<u>Subscripts</u>:	A,B - pertaining to solutes A and B, respectively; app - apparent value.

REFERENCES

1.   D. Merrick and J. Vernon, <u>Chem.</u> & <u>Ind.</u>, 1989 (6th
      Feb.), 55.
2.   *The Chemical Engineer*, 1988 (Oct.), 29.
3.   P. A. Ramachandran and M. M. Sharma, <u>Chem. Engng</u>
      <u>Sci.</u>, 1969, <u>24</u>, 1681.
4.   S. S. Bhagwat, J. B. Joshi and M. M. Sharma, <u>Chem.</u>
      <u>Engng Commun.</u>, 1987, <u>58</u>, 311.
5.   H. Takeuchi, K. Takahashi, T. Hoshino and M.
      Takahashi, <u>Chem. Engng Commun.</u>, 1980, <u>4</u>, 181.
6.   D. A. Dalrymple, T. W. Trofe and J. M. Evans, <u>Chem.</u>
      <u>Engng Prog.</u>, 1989 (Mar.), <u>85</u>, 43.
7.   L.C. Hardison, <u>Hydrocarbon Proc.</u>, 1985 (Apr.), <u>64</u>, 70.
8.   A. Mehra and M. M. Sharma, <u>Chem. Engng Sci.</u>, 1988,
      <u>43</u>, 1071.
9.   N. J. Suchak, K. R. Jethani and J. B. Joshi, 1989,
      submitted for publication.
10.   W. Gestrich, <u>Chem. Engng Technol.</u>, 1989, <u>12</u>, 33.
11.   *The Chemical Engineer*, 1989 (Apr.), 30.
12.   M. M. Sharma, <u>Chem. Engng Sci.</u>, 1983, <u>38</u>, 21.
13.   A. Mehra and M. M. Sharma, <u>Chem. Engng Sci.</u>, 1985,
      <u>40</u>, 2382.
14.   A. Mehra, <u>Chem. Engng Sci.</u>, 1988, <u>43</u>, 899.
15.   A. Mehra, A. Pandit and M. M. Sharma, <u>Chem. Engng</u>
      <u>Sci.</u>, 1988, <u>43</u>, 913.
16.   J. T. Tinge, K. Mencke and A. A. H. Drinkenburg,
      <u>Chem. Engng Sci.</u>, 1987, 1899.
17.   A. Mehra and M. M. Sharma, <u>Chem. Engng Sci.</u>, 1988,
      <u>43</u>, 2541.
18.   A. Mehra, <u>Chem. Engng Sci.</u>, 1989, <u>44</u>, 448.
19.   R. D. Holstvoogd, K. Ptasinski and W. P. M. van
      Swaaij, <u>Chem. Engng Sci.</u>, 1986, <u>41</u>, 867.
20.   Y. Tung, E. W. Grohse and F. B. Hill, <u>A. I. Ch. E.</u>
      <u>J.</u>, 1986, <u>32</u>, 1821.
21.   J. Szejtli, 'Inclusion Compounds' (eds. J.L. Atwood,
      J. E. D. Davies and D.D. MacNicol), Academic Press,
      London, 1984 (<u>3</u>), p.331.
22.   T. Sato, I. Toyoda, Y. Yamamori, T. Yonemoto, H. Kato
      and T. Tadaki, <u>J. Chem. Engng Japan</u>, 1988, <u>21</u>, 192.
23.   A. V. Patwardhan and M . M. Sharma, <u>Ind. Engng Chem.</u>
      <u>Res.</u>, 1989, <u>28</u>, 5.
24.   Z. Liu, J. W. Tierney, Y. T. Shah and I. Wender, <u>Fuel</u>
      <u>Proc. Technol.</u>, 1988, <u>18</u>, 185.
25.   Z. Bunder, W. Kotowski, B. Morawiee, J. Sitkiewiez,
      R. Marawski and F. Wanecki, <u>Polish Patent</u> No. 142059,
      1988, *cf.*, <u>Chem. Abstr.</u>, 1989, <u>111</u>, 7981.
26.   W. S. W. Ho, G. Doyle, D. W. Savage and R. L. Pruett,
      <u>Ind. Engng Chem. Res.</u>, 1988, <u>27</u>, 334.
27.   H. Hirai, S. Hara and M. Komiyama, <u>Chem. Letters</u>,

1986, **2**, 257.
28. J. D. Chase and B. B. Galvez, <u>Hydrocarbon Proc.</u>, 1981 (Mar.), <u>60</u>, 89.
29. A. A. Patwardhan and M. M. Sharma, 1989, submitted for publication.
30. A. Coenen, K. Kurt and E. Wienhoefer, <u>Ger. Offen.</u>, 836580, 1980, *cf.* <u>Chem. Abstr.</u>, 1980, <u>93</u>, 28693.
31. R. N. Lahiri, G. D. Yadav and M. M. Sharma, <u>Chem. Engng Sci.</u>, 1983, <u>38</u>, 1119.
32. J. A. Shropshire and D. J. Eustace, <u>U. S. Patent</u> No. 4124693, 1978.
33. B. R. Dunbobbin and W. R. Brown, <u>Gas Sep. & Pur.</u>, 1987, <u>1</u>, 23.
34. M. M. Sharma, <u>Trans. Faraday Soc.</u>, 1965, <u>61</u>, 681.
35. S. Chaudhuri and M. M. Sharma, <u>Ind. Engng Chem. Res.</u>, 1989, <u>28</u>, 870.
36. K. Chandrasekaran and M. M. Sharma, <u>Chem. Engng Sci.</u>, 1977, <u>32</u>, 275.
37. E. Sada, H. Kumazawa and S. Hattori, <u>Chem. Engng Commun.</u>, 1987, <u>57</u>, 95.
38. A. R. Nene, M. Tech. thesis, I. I. T., Bombay, 1979.
39. R. R. Bhave and K. K. Sirkar, <u>J. Membrane Sci.</u>, 1986, <u>27</u>, 41.
40. R. D. Noble, C. A. Koval and J. J. Pellegrino, <u>Chem. Engng Prog.</u>, 1989 (Mar.), <u>85</u>, 58.
41. M. Teramoto, H. Matsuyama, T. Yamashiro and Y. Katayama, <u>J. Chem. Engng Japan</u>, 1986, <u>19</u>, 419.
42. M. Teramoto, H. Matsuyama, T. Yamashiro and S. Okamoto, <u>J. Membrane Sci.</u>, 1989, <u>45</u>, 115.
43. D. R. Smith, R. J. Lander and J. A. Quinn, 'Recent Developments in Separation Science' (ed. N. N. Li), CRC Press, Cleveland, Ohio, 1977, Vol.3, PartB, p225.
44. E. Tsuchida, H. Nishide and H. Yoshioka, <u>Makromol. Chem., Rapid Commun.</u>, 1982, <u>3</u>, 693.
45. V. Kulkarni and R. Gobind, Paper presented at <u>Membrane Conference</u>, 1988, U. S. A.
46. *Air Products*, <u>U. S. Patent</u> No. 4761164, 1988, *cf.* <u>Chem. Abstr.</u>, 1989, <u>110</u>, 157019.
47. E. L. Cussler, R. Aris and A. Bhown, <u>J. Membrane Sci.</u>, 1989, <u>43</u>, 149.
48. H. Nishide and E. Tsuchida, Paper presented at <u>Membrane Conference</u>, 1988, U. S. A.
49. L. K. Doraiswamy and M. M. Sharma, 'Heterogeneous Reactions', Wiley, New York, 1984, Vol.2, Chapter 14.
50. D. O. Cooney and C. C. Jackson, <u>Chem. Engng Commun.</u>, 1989, <u>79</u>, 153.
51. Z. Qi and E. L. Cussler, <u>J. Membrane Sci.</u>, 1985, <u>24</u>, 43.

The Removal of Sulphur Dioxide from Flue Gases

W. S. Kyte

POWERGEN, HASLUCKS GREEN ROAD, SHIRLEY, SOLIHULL, WEST MIDLANDS B90 4PD, UK

1   INTRODUCTION

The reduction of the amount of pollutants emitted from the combustion of fossil fuels has become one of the largest growth areas in chemical separation processes. Most of the work in this field has taken place in an area far removed from the mainstream of chemistry and chemical engineering, namely, that of the production of electricity. This paper will describe methods of reducing sulphur emissions from the burning of fossil fuels to produce electricity.

2   REDUCTION OF SULPHUR DIOXIDE EMISSIONS

Acidification of the environment has been a growing concern over the last decade. In certain areas of Scandinavia and Scotland it has led to the loss of fish stocks in lakes. A major contribution to acidification of soils and waters is the emission of sulphur dioxide and nitrogen oxides from the combustion of fossil fuels.

The Central Electricity Generating Board (CEGB) has contributed significantly to the study of the acidification phenomenon, both by its own research and, in partnership with British Coal, by supporting financially a major study carried out by the Royal Society in collaboration with the Scandinavian Academies. This research will continue after privatisation in a major programme jointly funded by PowerGen and National Power, who will inherit the generating capacity of the CEGB.

Emissions of sulphur dioxide from UK power stations have fallen in recent years and there will be continuing significant reductions in emissions as the present generation of power stations is progressively replaced by plant incorporating "acid free" technology. Further reductions will ensue from measures to meet the requirements of the Large Combustion Plant Directive[1]. Each country in the EEC has agreed under this Directive to reduce its sulphur dioxide and nitrogen oxide emissions by various percentages of the 1980 emissions by specified dates. For the UK, the stipulations are that it should reduce its sulphur dioxide emissions from existing plant in three phases - 20% by 1993, 40% by 1998 and 60% by 2003.

The implications for the electricity supply industry in the UK are not clear but is is expected that the equivalent of 12 GW of generating plant will need to be fitted with sulphur dioxide removal equipment together with the use of some low sulphur fuels[2]. Since nearly all the electricity generated in the UK from non-nuclear sources is from coal, only emission reductions from the combustion of coal will be discussed.

It is not generally appreciated that there are many different ways of reducing sulphur dioxide emissions, some of the more relevant of which are summarised in Table 1. The most promising of these technologies are being actively investigated by PowerGen. Some of them, for example, the use of nuclear energy and higher thermal efficiencies for the conversion of coal to useful energy, have already had a substantial impact on sulphur dioxide emissions.

The largest wind turbine (1 MWe) ever constructed in mainland Britain is nearing completion on the PowerGen site at Richborough in Kent. In addition consent has been applied for the first UK windfarm at Capel Cynon in Wales. This is part of a wind energy programme jointly funded by the PowerGen and National Power Divisions of the CEGB together with the Department of Energy. This development programme could lead to 1000MW of windpower by the year 2005 providing about one per cent of the nation's electricity requirements.

TABLE 1: POSSIBLE METHODS OF REDUCING SO_2 EMISSIONS

- Energy conservation
- Increased thermal efficiency
- Alternative fuels
  - Low sulphur fuels
- Alternative sources of energy
  - Nuclear power
  - Solar power
  - Tidal power
  - Wave power
  - Wind power
  - Geothermal power
- Combined heat and power schemes
- Alternative combustion methods
  - Fluidised bed combustion
  - Gasification
- Removal of sulphur from fuel
- Removal of SO_2 from flue gases

Sulphur dioxide emissions from existing plant can be reduced either by removing sulphur from the coal before combustion or by removing sulphur dioxide during combustion or from the flue gases after combustion. Only those aspects which are relevant to large utility boilers will be discussed here.

3   COAL CLEANING

A detailed investigation[3] has been carried out to estimate the degree of extra sulphur removal that would be obtained if current washing processes were extended to include coals not presently treated. The study showed that there was only very limited potential for coal cleaning.

4 SULPHUR DIOXIDE REMOVAL DURING COMBUSTION

Sulphur dioxide can be removed during or just after combustion by injecting dry alkalis into the combustion gases. A detailed study[4] has shown that these systems would not be very competitive with wet processes for large utility boilers in the UK.

Fluidised bed combustion offers the prospect of both higher efficiency of coal conversion to electricity when operated in a combined cycle mode and lower emissions of acidic gases. Sulphur dioxide can be removed during combustion by the injection of limestone or dolomite. At small unit sizes fluidised bed combustion may be a better option than conventional plant fitted with flue gas desulphurisation.

5 FLUE GAS DESULPHURISATION (FGD)

The removal of sulphur dioxide from flue gases, whilst simple in the laboratory, is complicated in power plant by the combination of large scale and low concentration of sulphur dioxide. The scale and difficulty of the problem can be judged from the fact that a typical 2000 MW power station will generate some 10^7 m^3/h of flue gases which would have to be treated. In addition, the FGD plant will have to follow the operation of the power plant, conditions which are far removed from the steady state operations that are the ideal for chemical plant.

Modern methods of FGD began with the construction of various British plants (Battersea, Fulham, Bankside, North Wilford) during the period 1930-1960[5]. No further FGD plants were built in Britain due to the proven success of tall stacks in controlling ground level sulphur dioxide concentrations within acceptable limits at the points of maximum impact[6]. On the other hand, the early British work on FGD has been extended in the USA, Japan and W. Germany and there are now many FGD plants installed in these countries.

The use of flue gas desulphurisation on power stations substantially increases the cost of

electricity production and involves the utility
operator in technologies that lie in the realm of the
chemical industry. Before undertaking its FGD
programme, the CEGB carried out engineering and cost
studies of processes relevant to the UK situation in
order to gain familiarity with these technologies,
their costs and problems. The results of these studies
have been presented elsewhere[7-10].

     There are well over one hundred different flue gas
desulphurisation processes available in various degrees
of development, ranging from the laboratory scale to
full scale commercial units. Those processes that have
been developed to commercial scale are all capable of
removing 90% of the sulphur dioxide from the flue
gases[11].

Once-through Processes

     The first FGD units installed in the world, at
Battersea and Bankside, were once-through processes
which made use of the natural alkalinity of the River
Thames. The river water was passed through the FGD
plant and then, after suitable processing, discharged
back to the river. The only other process in this
category is one which uses seawater as the scrubbing
liquid. However, for a modern large power station the
natural alkalinity of the cooling water would not be
sufficient and addition of extra lime or limestones or
more seawater would be required. The dispersion of
effluents from the seawater process in estuaries or the
sea, and the effects of the effluents on marine life,
have not been fully elucidated. Further work is being
carried out on this process which is potentially the
cheapest.

Lime/Limestone Processes

     In these processes the alkaline absorbent liquor
is recirculated in a closed loop. The sulphated
reaction products are separated off and fresh absorbent
is added. The only water usage is to cover losses by
evaporation and that entrained with the product. A
purge is sometimes required to control impurity levels.
A quencher is used to cool the flue gases to avoid
concentration changes in the absorber due to
evaporation.

Lime or Limestone/Sludge Processes (Fig 1)

The vast majority of scrubbing systems installed in the United States use limestone as the alkali to neutralise sulphur dioxide. The resulting reaction product, a calcium sulphite/sulphate mixture, is then disposed of in ponds or disused mines. This mixture is very difficult to de-water and forms a thixotropic solid commonly known as 'sludge'. Fixation methods, including mixing with fly ash, cement, lime, soil etc. are often used to stabilise the sludge. The costs of disposal add significantly to the overall cost of the process. Such a process is considered to be unacceptable for use in the UK.

Gypsum Processes (Fig 2)

Since the disposal of sulphite/sulphate sludges is difficult there has been an incentive to produce a more amenable product. Gypsum ($CaSO_4.2H_2O$) is such a material as it can be sold for the manufacture of plasterboard, bag plaster or as a cement setting retardant. Gypsum can also be used for landfill if suitable disposal sites are available.

The lime/limestone process can be made to produce gypsum by the addition of an oxidation step in which calcium sulphite is oxidised to gypsum by air injection. Oxidation is usually carried out at a lower pH than that used for absorption. Intentional oxidation can also be carried out in the scrubber loop. Some of the processes produce gypsum of marketable quality and only these will be considered for use in the UK. These processes are the most common in use worldwide except in the US and even there some plant is being converted to produce gypsum.

Regenerative Processes

In these systems the absorbent is chemically or thermally regenerated for re-use and a saleable product (liquefied SO_2, sulphur or sulphuric acid) is generated. Ideally, there should be no disposal problems but all the processes produce some by-products which means that there is still a need for absorbent make-up and for by-product disposal.

Wellman-Lord Process (Fig 3)

The Wellman-Lord system employs absorption of

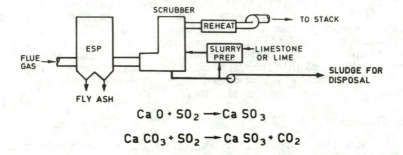

$$Ca\,O + SO_2 \longrightarrow Ca\,SO_3$$

$$Ca\,CO_3 + SO_2 \longrightarrow Ca\,SO_3 + CO_2$$

FIG. 1 LIMESTONE & LIME SLUDGE PROCESSES

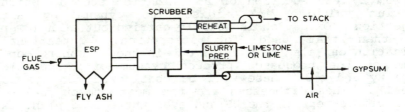

$$Ca\,O + SO_2 \longrightarrow Ca\,SO_3$$
$$Ca\,CO_3 + SO_2 \longrightarrow Ca\,SO_3 + CO_2$$
$$Ca\,SO_3 + \tfrac{1}{2}O_2 \longrightarrow Ca\,SO_4$$

FIG. 2 LIMESTONE & LIME GYPSUM PROCESSES

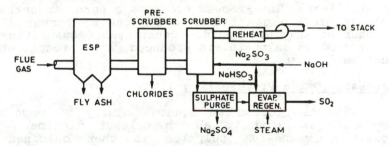

$$Na_2SO_3 + SO_2 + H_2O \rightleftharpoons 2\,Na\,HSO_3$$

FIG. 3 WELLMAN-LORD PROCESS

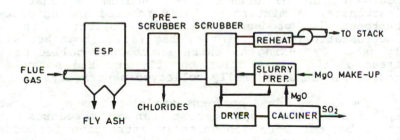

$$Mg\,O + SO_2 \rightarrow Mg\,SO_3$$

$$Mg\,SO_3 \rightarrow MgO + SO_2$$

FIG. 4 MAGNESIUM OXIDE PROCESS

sulphur dioxide in a concentrated solution of sodium
sulphite and regeneration of the resultant bisulphite
rich solution by steam stripping in an evaporator-
crystalliser. The process produces a pure concentrated
sulphur dioxide gas stream which can be converted to
sulphur or sulphuric acid. Considerable quantities of
impure sodium sulphate are produced, a by-product which
must be disposed of.

Magnesium Oxide Process (Fig 4)

     In this process an aqueous slurry of magnesium
hydroxide is used to absorb sulphur dioxide. The
resulting magnesium sulphite is then calcined to
release sulphur dioxide and to regenerate the
absorbent. The sulphur dioxide produced is available
for conversion to either sulphuric acid or sulphur.

Spray Drying Processes (Fig 5)

     During the past few years, the use of spray driers
for the absorption of sulphur dioxide has attracted
increasing attention. The spray drier in effect
replaces the absorber in a wet system, but is situated
upstream of the dust collector. The sulphur dioxide
absorbent can be either lime, sodium carbonate or
bicarbonate. A mixture of fully reacted and unreacted
absorbent is produced which is collected in an
electrostatic precipitator or bag filter together with
fly ash. By adding an extra small dust collector up
stream of the spray drier the fly ash and spray dry
product can be essentially collected separately.

     The spray dry process tends to have a lower
capital but higher running cost than wet processes, and
hence, becomes more economically attractive for older
plant or plant with a low load factor. It may also be
more suitable for smaller plant where the disposal of
the product may not be so difficult.

     6   CHOICE OF FGD SYSTEMS AND STATION IN THE UK

There are several factors, which are unique to the UK
electricity generation system which have had a marked
influence on the way in which FGD will develop in the
UK, namely,

     i) The bulk of generation is centred on large
         coal-fired power stations.

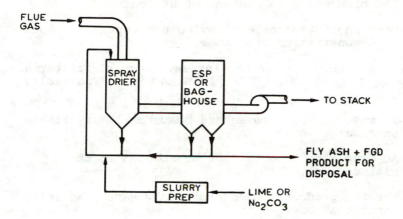

FIG. 5 SPRAY-DRY PROCESS

ii)   All units on the same site are of similar age.

iii) The high chloride content of UK coals.

iv)   Very short timescale allowing no scope for pilot
      or demonstration programme.

      In addition to these issues, several criteria
determine the type of FGD plant which is suitable,
namely,

  i) It must be well proven and commercially available
      on a large scale.

 ii) Adverse side effects must be reduced to the
      minimum.

iii) The FGD product should preferably have a socially
      beneficial use.

      When these factors and criteria are taken into
account, the only suitable processes are the
limestone/gypsum, regenerative, spray-dry and seawater
systems. Of these, only the two former are suitable
for the present FGD programme. The seawater process
may, after development, be suitable for a later new
coastal or estuarine site whilst spray-dry is best
suited to low load factor or smaller plant.

      For new stations, site specific factors can affect
the choice of the FGD system. For retrofit purposes,
the choice of FGD system and station are inter-related
and may need to be selected together. Site specific
factors may also be influenced by overall global
factors such as the marketing and disposal of FGD
products.

      The choice of stations for retrofit is governed by
many parameters among which the most significant are:-

  i) Sulphur levels in coal and coal burn.

 ii) Life of station and load factor.

iii) Ease of fitting of FGD.

 iv) Local product marketability and availability of
      disposal sites.

  v) Environmental factors.

vi) Supply of raw materials.

vii) Costs and economics.

The first FGD retrofit will be installed at Drax Power Station in Yorkshire which will be fitted with a limestone/gypsum process producing commercial quality gypsum for sale to the wallboard industry. These factors will be used to determine the choice of sites for the rest of the FGD programme.

7 OPTIONS FOR ADVANCED NEW PLANT

The use of natural gas in a gas turbine coupled with a steam turbine in combined cycles gives both increased efficiency and sulphur free emissions. As has already been mentioned, fluidised bed combustion may be a good option for smaller sized plant.

A recent study[12] has shown that coal gasification combined cycles offer potential in the future for generating electricity more cheaply with lower emissions to the atmosphere.

8 ACKNOWLEDGEMENT

This paper is published by permission of PowerGen which is currently a division of the Central Electricity Generating Board.

9 REFERENCES

1   Official Journal of the European Communities, 1988, L336

2   Kyte W S, 1988, 'A Programme for Reducing SO_2 Emissions from UK Power Stations - Present and Future', First Combined FGD and Dry SO_2 Control Symposium, EPA,EPRI, St. Louis, Oct 25-28, 1988

3   Jenkinson D E, 1985, 'Coal Preparation into the 1990's, Colliery Guardian, 301, July 1985

4   Burdett N A, Cooper J R P, Dearnley S, Kyte W S, Tunnicliffe M F, 1985, 'The Application of Direct Limestone Injection to UK Power Stations', Jl. Inst. Energy, LVIII, 435, 64, 1985

5   Kyte W S, Bettelheim J, Littler A, 1981, 'Fifty Years Experience of Flue Gas Desulphurisation at Power Stations in the United Kingdom', Chem. Engineer, <u>369</u>, 275, 1981

6   Barrett G W, Clean Air, 1979, <u>4</u>, 119

7   Bettelheim J, Cooper J R P, Kyte W S, Rowlands D T H, 1982, 'The Integration of a Regenerable Flue Gas Desulphurisation Plant onto a 2000 MW Coal Fired Power Station Site in the United Kingdom', I.ChemE Symp. Ser. 72, 1981

8   Kyte W S, Bettelheim J, Cooper J R P, 1982, 'Possible Fossil Fuel Developments within the Electric Power Generation Industry and Their Impact on other Industries', IChemE Symp, Ser., <u>78</u>, 1982

9   Kyte W S , Bettelheim J, Cooper J R P, 1983, 'Sulphur Oxides Control Options in the UK Electric Power Generation Industry', IChemE Symp. Ser. <u>77</u>, 1983

10   Kyte W S, Cooper J R P, 1984, 'The Disposal of Products from Flue Gas Desulphurisation Processes', 2nd Int. Conf. on Ash Technology, London, Sept 16-21, 1984

11   Kyte W S, 1989, 'Keynote Address: Technologies for the Removal of SO_2 from Coal Combustion', IChemE Symp. Ser. 160, 1989

12   Department of Energy, 1988, 'Prospects for the Use of Advanced Coal Based Power Generation Plant in the United Kingdom', Energy Paper No. 56, HMSO, London, (July 1988).